HEAT TRANSFER OF FINNED TUBE BUNDLES IN CROSSFLOW

Experimental and Applied Heat Transfer Guide Books

A. Žukauskas, *Editor*

A. Žukauskas and *J. Žiugžda*, Heat Transfer of a Cylinder in Crossflow
J. Vilemas, B. Česna, and *V. Survila,* Heat Transfer in Gas-Cooled Annular Channels
M. Tamonis, Radiation and Combined Heat Transfer in Channels
A. Žukauskas and *A. Šlančiauskas,* Heat Transfer in Turbulent Fluid Flows
J. Stasiulevičius and *A. Skrinska,* Heat Transfer of Finned Tube Bundles in Crossflow

IN PREPARATION

A. Žukauskas, V. Katinas, and *R. Ulinskas,* Vibration and Fluid Dynamics of Tube Bundles in Crossflow
A. Žukauskas and *R. Ulinskas,* Heat Transfer in Tube Banks in Crossflow

HEAT TRANSFER OF FINNED TUBE BUNDLES IN CROSSFLOW

J. Stasiulevičius
A. Skrinska

Institute of Physical and Technical Problems of Energetics
Kaunas, Lithuanian SSR

Edited by

A. Žukauskas

Academy of Sciences of the Lithuanian SSR, Vilnius

English-language editor

G. F. Hewitt

Thermal Hydraulics Division, AERE Harwell and
Imperial College of Science and Technology, London, England

◉HEMISPHERE PUBLISHING CORPORATION
Washington   New York   London

DISTRIBUTION OUTSIDE NORTH AMERICA

SPRINGER–VERLAG
Berlin   Heidelberg   New York   Tokyo

HEAT TRANSFER OF FINNED TUBE BUNDLES IN CROSSFLOW

Copyright © 1988 by Hemisphere Publishing Corporation. All rights reserved. Printed in the United States of America. Except as permitted under the United States Copyright Act of 1976, no part of this publication may be reproduced or distributed in any form or by an means, or stored in a data base or retrieval system, without the prior written permission of the publisher. Originally published by Mintis, Vilnius, USSR as, Teplootdacha poperechno obtekaemykh puchkov rebristykh trub

1 2 3 4 5 6 7 8 9 0   B C B C   8 9 8 7

This book was set in Times Roman by Hemisphere Publishing Corporation. The editor was Victoria Danahy; the production supervisor was Peggy M. Rote; and the typesetter was Cynthia B. Mynhier.
BookCrafters, Inc. was printer and binder.

Library of Congress Cataloging in Publication Data

Stasiulevičius, J.
  Heat transfer of finned tube bundles in crossflow.

  (Experimental and applied heat transfer guide books)
  Translation of: teplootdacha poperechno obtekaemykh puchkov rebristykh trub.
  Includes index.
  1. Heat—Transmission. 2. Tubes—Fluid dynamics. I. Skrinska, A. II. Žukauskas, A., 1923- III. Hewitt, G. F. (Geoffrey Frederick) IV. Title. V. Series.
TJ260.S73413 1987   621.402'5   87-17732
ISBN 0-89116-360-3 Hemisphere Publishing Corporation

DISTRIBUTION OUTSIDE NORTH AMERICA:
ISBN 3-540-18211-X Springer-Verlag Berlin

CONTENTS

Preface ix
Nomenclature xi

1 INTRODUCTION 1

1.1. Background 1
1.2. Early Studies and Development (Circular Fins) 2
1.3. Studies of a Spirally Finned System 9
1.4. Studies of Heat Transfer Mechanisms and Row-To-Row Variations 12
1.5. Hydraulic Drag and Pressure Drop 14
1.6. Concluding Remarks 17

2 CONDUCTION OF HEAT IN A FINNED TUBE 19

2.1. Fin Design 19
2.2. Temperature Gradient in a Circumferential Fin 22
2.3. Heat Dissipation from a Circumferential Fin 27
2.4. Heat Conduction of Spiral Fins 31
2.5. Fin Effectiveness 34
2.6. Thermal Conductivity of the Tube Supporting the Fins 38

3 FLOW OVER AND HEAT TRANSFER OF FINNED TUBE BUNDLES 50

3.1. Factors Controlling the Transfer of Heat from a Finned Tube Bundle 50
3.2. Coefficient of Heat Transfer from a Finned Tube 53
3.3. Selection of the Principal Parameters 59

3.4.	Effect of Finned Tube and Coolant Thermal Conductivities on Heat Transfer from Finned Tube Bundles	69
3.5.	Other Forms of Test Data Reduction	71

4 EXPERIMENTAL TECHNIQUES 75

4.1.	The Experimental Arrangement	76
4.2.	Determination of Heat Transfer from a Bundle	80
4.3.	Determination of Flow Variables	85
4.4.	Remarks on the Experimental Work	87

5 PRESSURE DROP IN FINNED TUBE BUNDLES 90

5.1.	Preliminary Remarks	90
5.2.	Experimental Data on Pressure Drop in Tube Bundles	92
5.3.	Analysis of Experimental Data	92

6 EXPERIMENTAL DETERMINATION OF THE DISTRIBUTION OF THE HEAT TRANSFER COEFFICIENT OVER A FIN 105

6.1.	Statement of the Problem	105
6.2.	Methodological Remarks	114
6.3.	Experimental Results	115

7 MEAN HEAT TRANSFER COEFFICIENT FROM A FINNED TUBE IN A BUNDLE 118

7.1.	Results of Experimental Study of Bundles of Tubes with Various Fin Geometries	118
7.2.	Results of an Experimental Study of Bundles with Varying Geometries with Constant Fin Geometry	126
7.3.	Correlation of Experimental Data	133

8 LOCAL HEAT TRANSFER FROM A FINNED TUBE IN A BUNDLE 138

8.1.	Preliminary Remarks	138
8.2.	Experimental Results	143
8.3.	Comparison of Model Heating Methods	149

9 EFFECTIVENESS OF THE BUNDLES UNDER STUDY 153

9.1.	Preliminary Remarks	153
9.2.	Comparison of Bundle Effectiveness on the Basis of the Surface Extension Factor	155
9.3.	Comparison of the Thermal Effectiveness of Bundles	157

9.4.	Overall Dimensions and Weight of Bundles	160
9.5.	Effect of Finning Geometry on the Thermal Performance, Overall Size, and Weight of Bundles	163

10 COMPARISON OF DATA AND PRACTICAL RECOMMENDATIONS 171

10.1.	Comparison of Data on Heat Transfer from Finned Tube Bundles	171
10.2.	Comparison of Data for the Euler Number for Finned Tubes	174
10.3.	Design Recommendations	176
10.4.	Concluding Remarks	182

APPENDICES 185

1.	Physical Properties of Air	185
2.	Experimental Data on the Hydraulic Drag of Staggered Bundles of Tubes With Spiral Fins in Crossflow Air	186
3.	Experimental Data on the Mean Heat Transfer from Staggered Bundles of Tubes with Spiral Fins in Crossflow Air	196
4.	Experimental Data on Local Heat Transfer from a Finned Tube with $d = 32$ mm, s from 4 to 7 mm, $h = 13.5$ mm in Crossflow Air	208
5.	Experimental Data for Determination of the Coefficient of Nonuniformity of Heat Transfer over a Finned Tube with $d = 32$ mm; $s = 6$ mm and $h = 9$ mm in a Staggered Bundle	213

REFERENCES 215

INDEX 221

PREFACE

The present book is the sixth in a series concerned with studies in thermophysics performed at the Institute of Physical and Technical Problems of Energetics of the Academy of Science of the Lithuanian SSR.

In addition to the general problems of convective heat transfer, which were dealt with in the previously published monographs, we, at the Institute, have investigated (over a period of several years) a number of practical problems, related to heat transfer and hydraulic drag bundles of finned tubes in crossflow of air at Reynolds numbers from 10^4 to 10^6.

This book is mainly written on the basis of studies performed by the authors, which were collected and methodically systematized into a single entity. It also presents, however, a survey of the principal literature sources dealing with this problem area.

The book begins by considering the thermal conduction behavior of a finned tube, and then goes on to consider the convective heat transfer behavior of finned tubes in crossflow. Data are presented on the hydraulic drag, on the nonuniform distribution of the heat transfer coefficient over the fin surface, and on mean and local heat transfer coefficients in finned tube bundles. The bulk of the experimental data is presented in the Appendix in tabulated form.

The initiative for, and performance of, these investigations of finned tube heat transfer are due to J. K. Stasiulevičius. Untimely death has cut short his life and scientific activity. He devoted many years of his life to the study of convective heat transfer; his work on heat transfer of bundles of smooth tubes in air flow is well known. J. K. Stasiulevičius completed this monograph while already seriously ill.

We wish to express heartfelt thanks to all the staff members who participated in performing the studies, and also to all those who assisted in bringing about the publication of this book.

A. Žukauskas
A. Skrinska

NOMENCLATURE

$a = s_1/d$		relative pitch of tubes in bundle
A_0, A_1, A_2		constants, Eq. (2.78)
$b = s_2/d$		relative longitudinal pitch of tubes in bundle
B		heat exchanger weight factor, kg/m²
c_p		specific heat capacity, J/kg · K
C_1, C_2		constants of integration, Eq. (2.14)
C_3, C_4, C_5		constants, Eq. (2.28–2.30)
d		diameter of base tube (fin root diameter), m
d_{eq}		equivalent diameter, m, Eq. (3.23)
d_F		equal surface diameter, m, Eq. (3.29)
d^*		duct diameter, m
D		fin diameter, m
E		fin effectiveness
E_0		energy factor, Eq. (9.5) and (9.8)
F		heat transfer surface, total outside surface of finned tube, m²
F_1		heat transfer surface of smooth tube, m²
f		cross-sectional area, m²
f_0		cross-sectional area of fin base, m²
G		weight of heat transfer surface, kg
h		fin height, m
h'		reduced fin height, m
I_0, I_1		zero and first row Bessel functions of imaginary argument
k		overall heat transfer coefficient, W/m² · K
K_0, K_1		zero and first row Bessel functions of imaginary argument
k_z		correction factor for the number of longitudinal rows of tubes in a bundle, Eq. (10.3) and 10.4)
k_s		correction factor for the geometrical position of tubes in a bundle, Eq. (10.4)
l		tube or duct length, m
l'		mean particle path, m, Eq. (3.25)
m		exponent of Reynolds number, Eq. (1.1)
N_0		circulation power per unit area, W
p		pressure, N/m²
Q		heat rate, W
q		heat flux density, W/m²
r		radial distance
r		exponent of Reynolds number, equation (1.15)
R		external fin radius, m
R_0–R_4		constants, Eq. (3.36)
s		fin pitch, m
s_1		transverse pitch of tubes in a bundle, m
s_2		longitudinal pitch of tubes in a bundle, m

xii NOMENCLATURE

	s_2'	diagonal pitch of tubes in a bundle, m
	t	temperature, °C
	T	absolute temperature, K
$u = s - \delta$		distance between fins, m
	U	wetted perimeter, m
	V	volume of heated surface, m³
	w	flow velocity, m/s
	w_0	bulk velocity, m/s
	w_y	velocity in most constrained cross section, m/s
	w_f	flow velocity between fin tips, m/s
	x	the wall, distance along the wall
$X = \beta h$		fin parameter
	z	distance along the tube axis, m
	z	number of longitudinal rows of tubes in the bundle
	α	heat transfer coefficient of fin, convective heat transfer coefficient of a finned tube, W/m² · K
	α_0	heat transfer coefficient on the surface of the finned tube between the fins, W/m² · K
	α_1	heat transfer coefficient of a smooth tube, W/m² · K
	α_{red}	reduced heat transfer coefficient of a finned tube, Eq. (3.6), W/m² · K
	α^*	conditional heat transfer coefficient of the finned tube in the absence of fins, Eq. (2.42), W/m² · K
	$\bar{\alpha}$	equivalent heat transfer coefficient, Eq. (2.97), W/m² · K
	$\bar{\alpha}h$	heat transfer coefficient averaged over the fin height, W/m² · K
	$\bar{\alpha}\varphi$	heat transfer coefficient averaged over the fin height and in tube perimeter, W/m² · K
	$\bar{\alpha}_m$	heat transfer coefficient averaged over the fin height, the tube perimeter, and for different fin pitches, W/m² · K
$\beta h = \sqrt{\dfrac{2\alpha}{\lambda_w \delta}} h$		fin parameter, dimensionless fin height
	δ	boundary layer thickness, m
	δ	fin thickness, m
	Δp	pressure drop, N/m²
	ϑ	dimensionless temperature
$\xi = 2\Delta p/\rho w^2$		drag coefficient
	ξ	factor for correcting the trapezoidal shape of the fin (Fig. 3.3)
	Π	heat exchanger compactness factor, m²/m³
	λ	thermal conductivity, W/m · K
	Λ	analogy coefficient
	μ	dynamic viscosity, N · s/m²
	γ	kinematic viscosity, m²/s
	ρ	density, kg/m³

$\varphi = \dfrac{F}{F_1}$ surface extension factor

ψ factor for nonuniformity of heat transfer on a fin surface

ψ_i heat transfer efficiency

$\mathrm{Nu} = \dfrac{\alpha d}{\lambda}$ Nusselt number based on the convective heat transfer coefficient

$\mathrm{Re} = \dfrac{wd}{\gamma}$ Reynolds number

$\mathrm{Pr} = \dfrac{\gamma}{a}$ Prandtl number

$\mathrm{Pe} = \mathrm{Re} \cdot \mathrm{Pr}$ Peclet number

$\mathrm{Eu} = \dfrac{\Delta p}{\rho w^2}$ Euler number

$\mathrm{Bi} = \dfrac{\alpha \delta}{\lambda_w}$ Biot number

Subscripts

- f free stream
- fin fin surface
- w wall
- sm smooth base tube
- int internal surface of tube
- m mean value
- red dimension obtained on the basis of reduced heat transfer coefficient
- 1 dimension related to the surface temperature of the finned tube at the fin base
- 2 dimension related to the temperature of the fin tip.

CHAPTER
ONE
INTRODUCTION

1.1. BACKGROUND

Augmentation of heat transfer and reduction of the coolant pumping power consumption by devices in which heat transfer occurs are the twin goals in improving the design of convective heat transfer equipment. The achievement of these goals is of particular importance for gas heat exchangers, in which the heat fluxes are low due to the poor thermal conductivity of the gas. Hence, usually, gas heat exchangers are built with large heat transfer surfaces. Nevertheless, they are extensively used in a large number of industries. The field of application of gas heat exchangers has widened considerably during the past decade due to developments in power generation (particularly nuclear) and in other new technologies.

The search for effective and most compact surface shapes, and also for new methods of heat transfer augmentation in gas heat exchangers, has been pursued in a large number of studies. The tube bundle is the most extensively used design of heat transfer surface for both low and high density fluids. However, the feasibility of improving the compactness of this design is limited. The use of small diameter tubes is a proven method for improving the compactness, but technological difficulties and high cost impose serious limitations on this solution.

The use of extended surfaces, or fins, attached to tubes by various methods, is an effective measure for increasing the compactness of tube bundles. Hence bundles made of tubes with different types of fins are the most extensively used form of extended surfaces. A bundle of finned tubes is a compact and highly effective unit from the point of view of heat transfer, particularly when used in crossflow (the heat transfer coefficient of tubes in crossflow is always higher than for the flow of fluid at the same velocity inside the tubes). Finned tube bundles are frequently utilized in designing gas-liquid heat exchangers, the external surface of which (the 'gas side') is usually constructed in such a manner that, at a given coolant flowrate and for the required heat removal rate, the lightest and smallest possible heat exchanger is achieved. Consideration must also be given to providing sufficient mechanical strength in such a heat exchanger and also to minimizing production costs.

A large number of finned tube designs have evolved in order to meet the above requirements. The first finned tubes were produced by casting; they had rather thick fins and a large fin pitch. Later, tubes were produced with fins soldered onto the outer surface. Improvements in soldering and welding methods made it possible to construct fins from metal tape, wound in the form of a continuous spiral. Tubes with slit, wire, slat-type, etc., fins appeared.

The development of new, improved fin shapes and the technologies for their manufacture has been given considerable impetus by the growing demands of nuclear power generation. This progress in development was also significantly assisted by the emergence of effective technologies for rolling and drawing continuous spiral fins, which made it possible to obtain a finned surface without resorting to welding or soldering, thus eliminating the problem of thermal resistance at the contact between the fin and the tube.

Not all the suggested (or actually produced) types of finned tubes survived the test of time, and came into actual and widespread use. This book is concerned with heat transfer and drag solely for tubes with circular fins, and also with continuous spiral fins, both of which have wide industrial applications.

1.2. EARLY STUDIES AND DEVELOPMENT (CIRCULAR FINS)

Heat exchangers with finned tubes have been used for many years. It was established early on that there is a very important limitation in the use of fins; this arises because the temperature differential between the gas and the fin surface decreases as the tip of the fin is approached and, sometimes, when the fins are high, the performance of a large part of the fin surface is rather poor. It therefore became necessary to establish an optimum fin height as well as shape.

The first studies were, in fact, related to the solution of these problems, as well as to the development of a technique for calculating the heat transfer coefficient for a finned surface. In these early studies, most attention was given to the question of the propagation of heat over the body of the fin. Assuming a heat transfer coefficient is constant over the entire surface of the fin, Schmidt [1] suggested (as early as 1926) solutions for heat transfer from right triangular and circular fins of constant thickness, by neglecting heat transfer at the fin tip.

He established that the most effective fin should have a cross section defined by two parabolas. This assumption was refined subsequently by Eckert and Drake [2] and by Makhin with his coworkers [3], who found that the most effective fin should have a cross section defined by arcs of radius R, where $R = \lambda_w/\alpha$ and where λ_w is the fin thermal conductivity and α the heat transfer coefficient to the surrounding fluid.

An approximated method for rectangular, and trapezoidally shaped fins, with correction for transfer of heat from the fin tip, was suggested by Harper and Brown [4]. According to this method, it was recommended that the effective height of the fin be calculated by adding one-half of its thickness to the actual height: this effective height was then used in the simplified equations (mentioned above) for calculating the heat transfer without taking account of the fin tip. However, the suggestion by Bogaerts and Meyer [5] that heat transfer calculations for circular and trapezoidally shaped fins should be performed exactly, using the method of successive approximations, gained wide acceptance, though such methods are tedious for routine design calculations. A further evaluation of simplified methods was made by Il'in and Styrikovich [6]. On the basis of exact solutions for heat transfer from a circular fin of constant cross section, with heat transfer at the tip, a simplified technique was developed for calculating circular fins of constant as well as variable (trapezoidal) thickness. Using this technique, accurate correction factors may be introduced into the expression derived from the assumption of no heat transfer at the tip. Il'in and Styrikovich [6] also showed that these corrections can be approximated with sufficient accuracy by increasing the effective fin height by one-half of its thickness, in agreement with the finding of Harper and Brown [4] mentioned above.

A significant contribution to analytic calculation of the heat conduction of a fin was made by Gardner [7] who, in addition to presenting generalized equations for the temperature distribution over the fin, gave correlation equations for determining the effectiveness of various fin types.

Although the above solutions (based on heat conduction in the fins) made allowance for the effect of shape, thickness, height and fin material properties on the temperature and heat flux distributions within the fin, they were all based on the assumption of a constant heat transfer coefficient between the fin and the surrounding fluid. No allowance was made for the flow conditions over the fins and for the actual distribution of the heat transfer coefficient. This was notwithstanding the fact that the experimental data on heat transfer available at that time [5, etc.], obtained even for a single finned tube, showed that the heat transfer coefficient is nonuniform over the fin. Moreover, conditions for flow over a tube within a bundle and over a single tube are different.

Therefore, the distribution of the heat transfer rate over the finned surface and the methods used for averaging the heat transfer coefficient to obtain a mean value for use in calculations became of considerable importance. There was a need for data on the effect of fin geometry, and also of the arrangement of tubes within the bundle. The non-availability of this data severely limited the practical application of available solutions for calculating the heat transfer behavior of finned surfaces. Hence the study of heat transfer from finned surfaces had to be performed by experimental methods, which aimed to directly simulate the process under study and to obtain nondimensional equations from similarity analysis of the experimental data.

Antuf'yev and coworkers [8, 10, 11], Baklastov [9], Karasina and coworkers [12, 13], Kuznetsov [16], Barbarich and Kirpichnikov [22], Kays and London [23], Fastovskiy and Petrovskiy [24], Kuntysh and Iokhvedov [25], Fortesque and Hall [26], Ueda and Harade [27], Yudin and coworkers [28, 29], Berman [30], Brauer [31] and the authors of [15] present experimental results on heat transfer from finned tubes over a wide range of factors such as fin geometry (Fig. 1.1) for tubes in staggered or in-line bundles (Fig. 1.2). The experimental results are correlated in nondimensional form. The most extensively used form is

$$\mathrm{Nu} = c\,\mathrm{Re}^m \tag{1.1}$$

or

$$\mathrm{Nu} = c\,\mathrm{Re}^m\,\mathrm{Pr}^{1/3} \tag{1.2}$$

$$\mathrm{St} = c\,\mathrm{Re}^{m-1}\mathrm{Pr}^{-2/3} \tag{1.3}$$

The overall resistance to heat transfer between a fin and its surrounding fluid is composed of the resistance to convective heat transfer to the coolant flow over the fin and the resistance to heat conduction in the fin material itself. Hence, the exact determination of the heat transfer behavior of a fin is rather difficult experimentally; it would involve the detailed measurement of the temperature

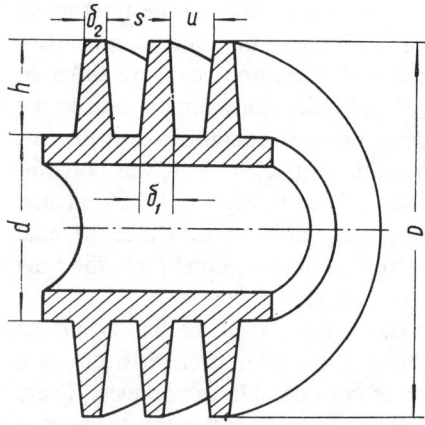

Figure 1.1 Section through a transversely finned tube.

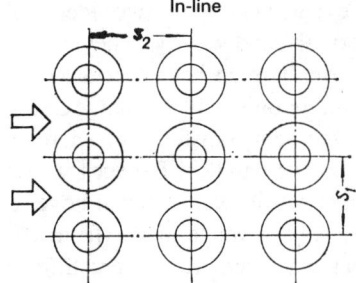 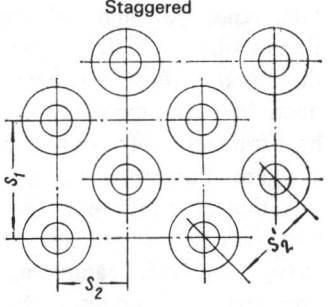

Figure 1.2 Arrangements of tubes in bundles.

field on the fin surface. However, usually, this difficulty is overcome by determining "conditional" heat transfer coefficients, in which the difference between the temperature of the flowing fluid and the temperature of the surface of the tube base (to which the fins are appended) serves as the temperature difference. The reference surface area is either the area of total surface, or, alternatively, of the surface of the tube to which the fins are appended. (In Soviet work the coefficient determined with this latter area is termed the "reduced coefficient," α_{red}.) This form of data reduction yields expressions which are relatively simple and convenient for use. It has the disadvantage that the correlations obtained cannot be generalized for all geometries and materials; it justifies itself, however, in determining the heat transfer characteristics of the specific finned surface for which the experiments were performed.

For studies of heat transfer to bundles of finned tubes, two main alternative techniques have been used. The first is the "complete thermal simulation technique" in which all the tubes in the bundle are heated. The second technique is the "local thermal simulation" method in which only one of the tubes is heated.

Data from extensive studies of heat transfer of bundles of machined steel tubes and of finned cast iron tubes ($d/s = 4.8$, h/s from 0.8 to 2.4, a from 1.5 to 2, b from 1.5 to 2) were published in 1948 by Antuf'yev and Beletskiy [8]. They employed both alternative thermal simulation techniques. The reference flow velocity was that in the most constrained flow passage. The reference dimension was the diameter of the finned tube and the heat transfer coefficient and physical properties were determined on the basis of tube temperature at the base of the fins.

Antuf'yev and Beletskiy [8] studied the row-to-row variation of the heat transfer coefficient; it was found that the coefficient became constant as early as in the third row of the staggered bundles and in the second row of the in-line bundles. The fin thickness was found to have little effect on heat transfer; within the limits of fin thickness under study, reducing the fin thickness by even as much as a factor of 5 resulted in a reduction of heat transfer by only 11 to 13%. This fact suggested that, for these experiments at least, the effect of thermal conductivity of the fin metal on heat transfer is insignificant. To check this, Antuf'yev and Beletskiy performed an experiment with two identical finned surfaces with materials of different thermal conductivity. They found that the difference between steel and aluminum fins (which have thermal conductivity (λ_w) differing by a factor of four) resulted in only a 10% change in the heat transfer coefficient. It was also found that the heat transfer coefficient in staggered bundles was somewhat higher than in in-line bundles. Except for low fins, this latter difference increases with h/d. These results were obtained over a wide range of fin geometries through a relatively narrow range of Re (from 10^4 to $6 \cdot 10^4$); both heating and cooling of the tubes was used. It is interesting to note that the direction of heat transfer, all other conditions remaining equal, resulted in up to a 20% difference in the heat transfer coefficient. This effect was allowed for by including a term T_f/T_w in the correlating equations for the cooling case. However, the conclusion regarding heat transfer direction was apparently erroneous and has not been confirmed in the large number of similar experiments which have been subsequently performed. On the basis of their studies, Antuf'yev and Beletskiy suggest the following optimal fin dimensions: fin height $h = 0.35\,d$, fin thickness $\delta \leq 0.035d$; they suggest that the distance between fins should be the minimum possible compatible with the operating conditions. Unfortunately, this large body of experimental data has not been used extensively, mainly because the original data presentation was on the basis of the reduced heat transfer coefficient.

Baklastov [9] investigated the heat transfer from a staggered bundle of copper tubes with circular fins ($d \times s \times h = 20 \times 4 \times 14$, $a = 2.6$ and $b = 2.2$) for Re $\leq 3.3 \lambda 10^4$; experiments were carried out with both cooled and heated tubes. The reference dimension used in data reduction was the diameter of the finned tube, the fluid physical properties being referred to the mean temperature of free air stream. The reference velocity was that at the minimum flow cross section within the bundle. Baklastov obtained the expression:

$$\mathrm{Nu}_{red} = 0.239\,\mathrm{Re}^{0.59} \qquad (1.4)$$

Antuf'yev [10] investigated a staggered bundle of copper tubes with circular fins ($d \times s \times h = 20 \times 3.6 \times 7.5$, $a = 1.8$ and $b = 1.6$) over approximately the same range of Re and, using the same data processing technique, obtained the formula

$$\mathrm{Nu}_{red} = 0.14\,\mathrm{Re}^{0.7} \qquad (1.5)$$

In another study [11] Antuf'yev investigated staggered bundles of tubes with spiral fins (d/s from 5.5 to 12.7, h/s from 0.87 to 2, $a \times b = 1.65 \times 1.42$), and correlated the experimental data on the basis of the surface extension factor ϕ. He found that:

$$\mathrm{Nu_{red}} = 0.33 \, \mathrm{Re}^{0.68} \tag{1.6}$$

Here the reference dimension was the fin diameter D, and the heat transfer coefficients were calculated on the basis of the surface area of a smooth tube whose diameter was the fin diameter; the remaining parameters were determined in the same manner as in [10].

Timofeyev and Karasina [12] present a correlation based on their experiments on heat transfer of in-line bundles of cast tubes with circular fins (d/s from 3.04 to 3.84, h/s from 1 to 1.4, $a \times b = 2 \times 2$), at $\mathrm{Re} \leq 7.1 \times 10^4$. This correlation had the form:

$$\mathrm{Nu_{red}} = 0.552 \left(\frac{h}{d}\right)^{0.57} \mathrm{Re}^{0.65} \tag{1.7}$$

the reference dimension here being the fin pitch. The physical properties were defined as those at the mean bulk temperature, and the air velocity was defined as that measured at the minimum flow cross section within the bundle. The heat transfer coefficients were calculated using an area corresponding to the entire outside surface of the finned tube.

It is seen from the above that the early experimental data were reduced by a large variety of techniques, which makes comparison of the results obtained by the different investigators very difficult.

Karasina, who, in 1952, investigated heat transfer from bundles with cast iron fins [13], made a distinction, in her data reduction, between the thermal resistance of the fin metal, obtained analytically, and the resistance to heat transfer between the fin surface and the fluid. This made it possible to find the relationship between the measured "reduced" coefficient (based on the base tube surface area and fin base temperature—see above) and the actual convective heat transfer coefficient between the tube outer surface and the fluid. This allowed Karasina to find a generalized technique for calculating the heat transfer from such tubes.

As was explained above, analytical solutions for conduction in fins are usually derived from the assumption that the convective coefficient is constant over the tube surface. By making measurements of both "reduced" and average actual convective coefficients, Karasina was able to obtain the value of a correction factor of the analytical conduction solution which makes implicit allowance for the nonuniformity of the heat transfer coefficient distribution over the fin surface.

Using this technique and data of Antuf'yev [8, 10], Timofeyev and Karasina [12] and Kuznetsov and Shcherbakov [14], made it possible to obtain the first correlations of actual convective heat transfer coefficients. These were for bundles of tubes with circular fins with d/s from 3 to 4.8, $a = b = 2$ and Re from 3×10^3 to $2 \times 5 \times 10^4$.

For staggered bundles the correlation obtained by Karasina [13] was:

$$\text{Nu} = 0.223 \left(\frac{d}{s}\right)^{-0.54} \left(\frac{h}{s}\right)^{-0.14} \text{Re}^{0.65} \tag{1.8}$$

and for in-line bundles:

$$\text{Nu} = 0.104 \left(\frac{d}{s}\right)^{-0.54} \left(\frac{h}{s}\right)^{-0.14} \text{Re}^{0.72} \tag{1.9}$$

The reference dimension in this correlation was the fin pitch s.

Equations (1.8) and (1.9) are used in the standard Soviet handbook [15] as working equations. Calculations using them were facilitated by constructing nomograms [15, 16].

Where high heat fluxes are required from finned surfaces, Reynolds numbers are encountered which are beyond the range of Equations (1.8) and (1.9). In this context, it is particularly important to consider the studies of Brauer [17]; these were performed at elevated air pressures with 15 in-line and staggered bundles of tubes with circular fins with Re from 2×10^3 to 5×10^5. Experiments with bundles of smooth tubes, performed in the same study, showed that dimensional similarities are also valid at elevated air pressures. Similar conclusions were also drawn in our studies with smooth tube bundles [18–20].

The experiments by Brauer were performed by the complete simulation technique, with all the tubes heated. The average heat flux used in determining the heat transfer coefficient was measured by determining the condensation rate of the steam used to heat the tube; in defining the coefficient, a surface area was used corresponding to smooth tubes whose diameter is equal to the fin base diameter. This diameter was also taken as the reference dimension in the correlations. Using experimental data on four-row in-line bundles at $a \simeq b \simeq 2$, h/d from 0.35 to 0.54, $s/d \simeq 0.2$, Brauer obtained the correlation

$$\text{Nu}_{\text{red}} = c \, \text{Re}^{0.63} \, \text{Pr}^{1/3} \tag{1.10}$$

where c is constant with a range from 0.248 to 0.714.

Experimental data for staggered bundles with $a = b \simeq 2$, h/d from 0.07 to 0.5 and $s/d \simeq 0.2$ to 0.27 were correlated in a similar manner; c ranged from 0.21 to 0.36 and the exponent Re ranged from 0.55 to 0.63.

Brauer additionally showed that, as the outer surface of tubes is progressively extended by using more fins, more and more tightly spaced, the added surface is less and less effective in increasing the heat transfer rate. This indicates that the boundary layer between the fins is affecting the heat transfer; the smaller the distance between the tubes, the thicker the boundary layer. Brauer concluded that a sparsely finned surface with widely spaced fins is more effective than surfaces with small fin spacings. Clearly, the more fins that are used, the more the pumping power required to pass the fluid through the bundle; Brauer found that the transmission of the same amount of heat at the same pumping power requires a 20–25% smaller heated surface in staggered than in in-line bundles. The results of these studies were included as a supplement to the standard Soviet heat engineering handbook [21].

1.3. STUDIES OF A SPIRALLY FINNED SYSTEM

The extensive practical use of heat exchangers with spiral fins, brought about by the great progress in manufacturing methods (see, among others, the paper by Barbatich and Kirpichnikov [22]) led to a need for appropriate, specific studies. It was assumed that the heat transfer coefficients for these tubes would be higher than those for tubes with circular fins, since the fluid turbulence level was expected to be higher; in the case of ordinary circumferential fins the flow moves parallel to the fin surfaces, whereas in the case of spiral fins it is diverted and stirred by the fins which are no longer precisely in the flow direction.

In 1958, Kays and London [23] published and correlated a large body of experimental results on a large variety of types of extended surfaces. Their data included results for a number of staggered bundles of bimetal tubes with circular fins (copper tube and aluminum-fin envelope) and heat transfer of copper tubes with integral copper spiral fins at Re from 10^2 to 10^4. The results were correlated using the form of Eq. (1.3). Comparison of the results for the two types of fin confirmed that bimetal tubes suffer from significant thermal resistance between the inner copper tube and outer aluminum fin envelope; this resulted in a preference for integrally finned tubes.

Fastovskiy and Petrovskiy [24] present results of a study similar to that of Kays and London. Their data were for six-row staggered bundles of copper and aluminum tubes with continuous spiral fins at Re $\leq 10^4$. Their experiments were performed with steam-heated tubes, with the steam condensing inside the tubes. The tube diameter at the fin base was used as the reference diameter, and the physical properties of the air were taken at the wall temperature. The heat transfer coefficients were of the "reduced" form, i.e., the surface area was estimated on the basis of "base tube area;" this corresponds to the area of a bundle of tubes whose diameter is that of the fin base, that bundle having the same number of tubes as the relevant finned tube bundle. The fins of the tubes under study had the following parameters: s/d from 0.17 to 0.27, h/d from 0.31 to 0.51, a from 1.6 to 2.2 and b from 1.4 to 1.9. Fastovskiy and Petrovskiy's results are described by the equation:

$$\text{Nu}_{\text{red}} = A \, \text{Re}^{0.643} \, \text{Pr}^{1/3} \qquad (1.11)$$

The values of A for the bundles under study (no correlation is presented) range from 0.75 to 2.36.

Kuntysh and Iokhvedov [25], who investigated five-row staggered bundles of steel, copper and brass tubes with spiral fins, drew some interesting conclusions. The Re range from 4×10^3 to 4×10^4. The air physical properties were taken at a temperature which was the arithmetic mean of the temperatures of air entering and leaving the bundle. In spite of the fact that the thermal conductivities of the three metals used were very different, the heat transfer coefficient for the respective metal surfaces did not differ greatly, all other conditions remaining equal: it was 1.1 higher for the brass and 1.26 higher for the copper, as compared with the steel tubes. This reconfirmed the fact that the principal resistance to heat transfer from a finned tube bundle is the thermal resistance between the fin surface and the air flow, which is primarily controlled by the stream flow conditions. The effect of thermal conductivity of fin material on heat transfer was correlated in [25] by the ratio λ_w/λ_f, where λ_w and λ_f are the wall (metal) and fluid thermal conductivities, respectively. The experimental data were correlated by the expression

$$\mathrm{Nu}_{red} = 0.0614 \varphi \eta \left(\frac{\lambda_w}{\lambda_f}\right)^{0.11} \mathrm{Re}^{0.64} \tag{1.12}$$

where ϕ is the surface extension factor, whereas η is a coefficient (the "conditional coefficient of bundle effectiveness") which allows for the fin conductance effect. A similar procedure was used by Fortescue and Hall [26], Ueda and Harade [27] and Yudin [28].

Yudin et al. [29] also present data on staggered bundles of tubes with spiral fins for $\mathrm{Re} \leq 5 \times 10^4$. Their studies were performed by the local simulation technique. The data were correlated by the method used by Karasina [13], described above. They estimated the heat transfer performance (measured in terms of the heat transfer rate) for the bundles under study as a function of the fin parameters. It was found that performance increased significantly with increasing fin height only for fins up to 10 mm high (at a constant fin pitch of $s = 5$ mm). Using higher fins ($h = 15$ mm) did not result in a significant increase in the heat transfer rate. The optimum fin dimensions were: $h = 10$ mm, $s = 3$ mm at $\delta = 1$ mm.

Berman [30] investigated staggered bundles of integrally drawn spiral-finned tubes at Re from 2×10^4 to 4×10^4 and appeared to detect a region with a higher heat transfer rate (depending on the fin parameters). He found that, in this region, the exponent Re in Eq. (1.1) increases to the range 0.97–1.227. Given the high value of the exponent, and also the fact that Berman obtained different heat transfer coefficients in the same bundles, depending on the direction of the heat flux, (heating and cooling), it may be inferred that errors had crept into the study.

The widest range of Re (from 5×10^3 to 10^5) in studies of heat transfer from tubes with spiral fins was covered by Brauer [31]. He investigated eight staggered and in-line bundles over a wide range of bundle, tube and fin geometries. The data obtained for staggered bundles were correlated in the form of Eq. (1.2). Brauer compared heat transfer coefficients for staggered and in-line tube bundles and found that the staggered arrangement was much more advantageous. He showed that, for a given pumping power, the staggered arrangement required 30% less heating surface than the in-line arrangement, consistent with his earlier studies mentioned above.

In all of the above-mentioned studies with spirally-finned tubes, the expected behavior relative to a bundle with circular fins was verified; the heat transfer coefficient was equal or, usually, higher than that for the circular fin case.

The principal objective in a large number of investigations of heat transfer from finned tube bundles is the selection of optimal finning parameters. The effect of tube location within a bundle has received less attention; however, tube location effects have been extensively investigated in studies with bare tubes over a wide range of Re and for different pitch arrangements. Thus, it was shown by Žukauskas and coworkers [32] that there exists a perceptible difference in the heat transfer behavior of tubes in different locations within the bundle. Differences in the heat transfer coefficient, as high as 30%, have been observed in bundles with large transverse spacing between the tubes (high a/b). Thus one might expect similar variations in the case of finned bundles. The limited studies of tube location effects will now be reviewed.

Jameson [33] investigated heat transfer in four-row staggered tube bundles with spiral fins ($d/s = 5$, h/s from 1.5 to 3) with Re from 4×10^3 to 6×10^4 and with transverse and longitudinal pitch ratios ($a = s_1/d$ and $b = s_2/d$) ranging from 1.9 to 3.5 and from 1.03 to 2.45, respectively. It was found that the tube location within the bundle has little effect on heat transfer. Broadly speaking, the same conclusion was drawn by Antuf'yev and Beletskiy [8]. On the other hand, Zozulya et al. [34] found, in studies of heat transfer in staggered bundles, that the location of the finned tubes within the bundle had a significant effect. Their experiments were performed for $a = 1.82$ to 2.41 and $b = 1.36$ to 1.89. The tube location effect is described in their correlation:

$$\mathrm{Nu_{red}} = 0{,}6115\,\mathrm{Re}^{0.65}\left(\frac{s_1}{d}\right)^{0.9}\left(\frac{s_2}{d}\right)^{-0.4} \tag{1.13}$$

where the hydraulic equivalent diameter at the minimum cross section is used as the reference dimension. In these studies, heat transfer coefficients were calculated on the basis of the total surface of the finned tube. The studies were performed with 10 staggered bundles of aluminum tubes with continuous spiral fins ($d \times s \times h = 22 \times 3.3 \times 8$) at Re $\leq 2 \times 10^4$. The experiments were performed by the local simulation technique.

In pursuance of the optimization of bundle configurations, a very wide range of studies has been performed. For instance, Yudin et al. [35] investigated 17 bundles of tubes with spiral fins ($d \times s \times h = 32 \times 6 \times 9$). They found that with $a = 1.7$ to 3, $b = 1.2$ to 3 and Re from 10^4 to 5×10^4 the heat transfer coefficient varied, for staggered tubes, by as much as 60%. For a given value of Re, it was established that heat transfer is affected primarily by the transverse pitch; heat transfer rates increase with the pitch. Heat transfer coefficients for in-line bundles increased with the longitudinal pitch and were virtually independent of the transverse pitch. A staggered bundle with a $a = 3$ and $b = 1.2$ was selected as the optimum from the point of view of tube configuration.

Mirkovič [36] investigated 13 staggered bundles with spiral fins at Re from 3×10^3 to 5×10^4 to determine the effect of fin geometry and bundle configuration on heat transfer. The equation

$$\mathrm{Nu}_{\mathrm{red}} = 0.224 \left(\frac{s_1-d}{d}\right)^{0.1} \left(\frac{s_2-d}{d}\right)^{-0.15} \left(\frac{1-s\delta}{sh}\right)^{0.25} \mathrm{Re}^{0.662} \mathrm{Pr}^{0.33} \quad (1.14)$$

obtained by Mirkovič for s/d from 0.083 to 0.17, h/d from 0.18 to 0.67, a from 4 to 4.75 and b from 2.37 to 3.14, indicates that these variables have little effect on heat transfer from the bundle.

Schmidt [37] correlated a large body of his own and published data on heat transfer at Re from 10^3 to 5×10^4. In the nondimensional equations suggested by him for those tube locations within the bundle for which experimental data were available, the effect of finning geometry was estimated by an exponent to the surface extension factor.

Yudin et al. [38] correlated the experimental data on heat transfer of bundles of finned tubes for Re $\leq 2.5 \times 10^4$ using the Karasina technique [13] described above.

1.4. STUDIES OF HEAT TRANSFER MECHANISMS AND ROW–TO–ROW VARIATIONS

Krischer and Kast [39] and also Grass and Coenen [40] were concerned with study of the mechanism of heat transfer from a single spiral-finned tube. As a result of analysis of data for tubes with circular fins, Krischer with Kast suggest that a relationship exist between the finning parameters, Reynolds number, and the velocity reduction in the space between the fins. Their design equations relate these quantities. They obtained nondimensional equations, which allow heat transfer from a finned tube to be treated not as a case of a body in external flow, but as a process occurring in the presence of flow in the channel formed by the space between the fins.

Grass and Coenen [40] investigated heat transfer from finned tubes with various fin profiles at Re from 10^4 to 5×10^4. Wishing to determine the reason for changes of the heat transfer coefficient of the fin as a function of Re and fin geometry, they concluded that basically this variation is a function of the properties of flow over the tube. In the case of higher fins the gas flow breaks up and gradually becomes a planar flow, which is less effective from the point of view of heat transfer. A boundary layer is formed on the fins whose thickness increases with fin length and which is detrimental to heat transfer. However, the effect of the boundary layer may be insignificant as long as the distance u between the fins is greater than twice the thickness of the boundary layer. In addition, Grass and Coenen showed that the performance of a heat exchanger can be significantly improved by proper selection of fins. At a given heat flux they were able to reduce the heat exchanger volume by 30%, while at same time reducing its pressure drop by 60%.

Lapin and Shurig [41] obtained an equation for heat transfer from staggered tube bundles with circular fins ($d \times s \times h = 16 \times 3.2 \times 11$ mm) at Re from 1.6×10^3 to 1.1×10^4. They found that the flow turbulence increases as the number z of rows is increased (from 1 to 8); this results in a rise in the heat transfer rate from row-to-row in the direction of flow. The constant in the correlating equation falls off gradually as the numbers of rows is increased, whereas the exponent of Re increases from 0.36 to 0.85. The heat transfer coefficient becomes nearly constant independent of the row numbers for the fourth and the subsequent rows.

Heat transfer in the front rows of finned tube bundles was investigated in detail by Hirschberg [42] and Hufschmidt [43], who also confirm that the heat transfer coefficient for a tube within the bundle becomes constant starting with the fourth row. It is rather interesting to examine the paper by Kast [44], who attempted to compile a universal graphical relationship which would correlate heat transfer from finned and bare tubes. He determined the optimum values of s_2/d for bundles with staggered and in-line tubes. He also obtained data on the effect of the number of longitudinal tube rows on heat transfer. Comparison of data confirmed the significant advantage of the staggered configuration.

The question of the correction factor, making allowance for nonuniformity of the distribution of the heat transfer coefficient over the fin surface (introduced by Karasina [13]), and its effect on the effectiveness of a finned tube was afforded a great deal of attention in later studies by Migay [45], Brauer [46] and Sasin [47]. The present authors addressed this problem in 1965-1966 [48-50].

Lymer and Ridal [51] employed the properties of thermal conductivity of the fin metal as $\lambda_w \to \infty$, when the fin effectiveness factor E tends to unity ($E \to 1$) and, respectively $\alpha_{red} \to \alpha$, to suggest a method for determining α_{red} by extrapolation, using experimental data obtained for a bundle of finned tubes with different fin-metal thermal conductivities (λ_w). This method was subsequently further developed by Yudin [28] and Pshenisnov and Lukhnov [52].

All the studies mentioned so far were concerned with experimental results for *mean* heat transfer coefficients. However, measurement of the *local* heat transfer coefficient allows a more profound insight into the mechanism of heat transfer and flow over finned surfaces. Lemon et al. [53] and McAdams et al. [54] investigated local coefficients of heat transfer over the circumference of air-cooled finned cylinders of aircraft engines, the cylinders being enclosed in a shell. Weiner, et al. [55], who investigated heat transfer with a single finned tube in crossflow, present experimental data on the distribution of the heat transfer coefficient over the fin surface.

Usually the temperature at the fin base is taken to be constant over the entire base cross section. Actually, the cross section of the fin base, over which heat is transmitted to or from the fin-bearing tube, is not at constant temperature. The effect of this factor on heat transfer from a finned tube was investigated analytically by Drischer with Kast [39]. The same problem was investigated experimentally by the electric analog technique by Comossa [56].

Neal and Hitchcock [57] investigated the local coefficients of heat transfer in three types of large-scale staggered finned tube bundles ($d \times s \times h = 154 \times 17 \times 41$ mm, a from 2.0 to 2.68, b from 1 to 1.74). They found that the flow conditions, local velocities and turbulence levels are different in different longitudinal rows of the bundle, and are a function of the height and circumference of the fin. This causes significant variation in the local heat transfer coefficients. At about the same time, we published our papers concerned with aspects of local heat transfer from a finned tube [49, 58].

1.5. HYDRAULIC DRAG AND PRESSURE DROP

A sensible selection of a heat exchanger-heat transfer surface is possible only by simultaneously considering the problem of heat transfer and *pressure drop* or *hydraulic drag*. For this reason many studies deal with both heat transfer and pressure drop or hydraulic drag.

The majority of previously described investigators were concerned with the effect of fin geometry on the hydraulic drag, and only a few of them address systematically the effect of tube location within the bundle. Experimental data for pressure drop Δp are often correlated in terms of Euler number Eu with expressions of the type

$$\text{Eu} = \frac{\Delta p}{\rho w^2} = k \text{Re}^{-r} \tag{1.15}$$

where ρ and w are the fluid density and velocity (defined in an appropriate way) respectively. Alternatively, the data may be correlated in terms of drag coefficient ξ, defined as:

$$\xi_0 = \frac{2\Delta p}{\rho w^2} \frac{1}{z} \tag{1.16}$$

where ξ is the number of ions.

The physical nature of drag (friction due to viscous shear forces, and form drag associated with boundary layer separation) in finned tube bundles has not been widely explored. Whereas the friction drag comprises only several percent of the total drag of a bundle of base tubes [32], a significantly higher proportion of drag is due to friction in finned tube bundles [17]. However, since in practice the objective of design calculations is to obtain the total drag of the bundle, it is unnecessary to estimate each part of the drag separately. Thus, usually, only the overall bundle drag is given in the aforementioned investigations.

Antuf'yev and Beletskiy [8], who investigated drag in six-row bundles of tubes with circular fins, confirmed that the drag coefficient of bundles is a power law function of Re, as given by Eq. (1.15). They found that self-similar flow starts in staggered bundles at Re $\geq 5 \cdot 10^4$. In in-line bundles this kind of flow occurred over the entire range of Re studied (10^4 to 10^5). In addition, they found that the drag coefficient referred to a single row decreased with increasing number of longitudinal rows, but this reduction is minimal when the number of rows increases beyond three. On the basis of this, they concluded that, in order to obtain constant drag coefficient values, a bundle must have at least three longitudinal rows. It was found that for in-line bundles with $a = b = 2$ and $h/d = 0.209$ and u/d from 0.167 to 5 the Euler number for a bundle can be obtained from the expression:

$$\mathrm{Eu} = 1.35 z \left(\frac{h}{d}\right)^{0.45} \left(\frac{u}{d}\right)^{-0.72} \mathrm{Re}^{-0.24} \qquad (1.17)$$

and the corresponding expression for an in-line bundle is given by the equation

$$\mathrm{Eu} = 0.094 z \left(\frac{h}{d}\right)^{0.5} \left(\frac{u}{d}\right)^{-0.58} \qquad (1.18)$$

The above equations illustrate the kind of effect exerted by the fin geometry on pressure drop in bundles. In more compact, staggered and in-line bundles the drag coefficients were found to be respectively proportional to $(h/d)^{0.2}$ and $(h/d)^{0.3}$. It was found experimentally that a five-fold increase in the relative fin thickness had no effect on bundle pressure drop.

Brauer [17, 32] performed extensive studies of pressure drop in bundles, made of tubes with circular and spiral fins, at high air pressure (Re $\leq 10^5$) and found that self-similar flow sets in only in certain in-line bundles (basically with dense fins, $u = 2$ mm) at Re $\simeq 8 \cdot 10^4$, setting in much later in staggered bundles. As did Antuf'yev and Beletskiy [8], Brauer found that the drag coefficient referred to a single tube decreases with increasing z and stabilizes at $z = 4$. The relative difference in the coefficient between a single row and four rows decreases with increasing Re. It was shown by comparing the drag of in-line and staggered bundles with similar fins that the drag of the latter is somewhat higher. Brauer does not present correlations of his data.

Data obtained by Kuntysh and Iokhvedov [25] for a five-row bundle were represented by these investigators in the form:

$$\mathrm{Eu} = 5.14 \varphi \eta \, \mathrm{Re}^{-0.28} \qquad (1.19)$$

where ϕ is the surface extension factor and η is a coefficient whose value depends on the fin geometry: for a trapezoidal fin with $h \times s \times \delta = 8 \times 3.25 \times 0.3$ mm, $\eta = 1$; for a rectangular fin cross section with $h \times s \times \delta = 10 \times 3.5 \times 0.6$ mm, $\eta = 0.8$; for a rectangular fin cross section with $h \times s \times \delta = 10 \times 2.5 \times 0.6$ mm, $\eta = 0.685$.

Kays and London [23] describe their experimental data using Eq. (1.15). For the six-row bundles under study and for Re from 6×10^3 to 6.2×10^4, the ranges of values of coefficient k and exponent r were: k from 2.5 to 5.32 and r from -0.25 to -0.29.

Yudin, et al. [29] found in investigating the pressure drop over an eight-row bundle ($a = 3$, $b = 1.6$) of tubes with $d \times s \times h = 23 \times 2.5 \times 10$, cooled by a number of fluids, that the Euler at $Re \leq 1.2 \times 10^5$ can be correlated by the equation

$$Eu = 31 \, Re^{-0.26} \qquad (1.20)$$

Self-similar flow sets at $Re > 1.2 \times 10^5$, and then the above expression becomes

$$Eu = 1.45 \qquad (1.21)$$

The reference dimension in the above studies is the diameter of the fin-bearing tube.

It is seen by examining Eqs. (1.19)–(1.21) how different finning parameters affect the pressure drop for the bundles. The effect of bundle configuration on pressure drop was not explored in these studies. Below, we now consider several studies in which the latter aspect is the primary concern.

Yudin et al. [29] found that the drag coefficient for bundles of staggered tubes with spiral fins increases with reduction in the transverse pitch, while being virtually independent of the longitudinal pitch. This is also confirmed in the book by Fastovskiy with Petrovskiy [24], where it was reported that increasing s_1/d from 1.04 to 1.15 was found to reduce the drag of a bundle by 13%.

Jameson [33] presents the equation

$$\xi = c \, Re^{n-2} z \qquad (1.22)$$

obtained by wind tunnel testing of eight-row staggered bundles at a from 1.9 to 3.6 and b from 1.1 to 2.5. The value of exponent n in the correlation ranged, depending on the bundle configuration, from 1.7 to 1.9. Bundles with denser tube packing had lower values of n. Jameson suggested that the mean value of n for the range of relative pitches under study be taken as 1.75. The variation in n as a function of the density of tube packing was attributed by Jameson to changes in the relationship between the velocity and pressure variations in the space between the tubes.

Yudin and Tokhtarova [35] present a correlation for the pressure drop for a six-row bundle, making allowance for the arrangement of tubes in a staggered bundle at $Re \leq 2 \times 10^4$:

$$Eu = 54 \, Re^{-0.3} \left(\frac{s_1}{d}\right)^{-0.2} \left(\frac{s_2}{d}\right)^{-0.85} \qquad (1.23)$$

The above expression was obtained from wind tunnel testing of ten bundles of tubes with spiral fins ($d \times s \times h = 22 \times 3.3 \times 8$) at a from 1.82 to 2.41 and b from 1.36 to 1.89. The optimal configurations were that in which $a > b$.

The experimental data cited above show that, in general, the aerodynamic drag of bundles increases with the heat transfer rate.

1.6. CONCLUDING REMARKS

To find the optimal type of a finned surface for given heat exchanger applications, it is frequently necessary to estimate the most effective type of fins and arrangement of finned tubes in the context of the specific application. This problem cannot be solved by simply comparing the available correlations of heat transfer and pressure drop for the finned tube bundles. One must have additional information relating to the dimensions and weight of the heat exchanger and of its cost. The latter is made up of the cost of manufacture, plus the cost of pumping the coolant through the finned tube bundles.

Methods for estimating the effectiveness of surfaces were developed by Kays with London [23], Fastovskiy with Petrovskiy [24] and Kirpichev [65]. They were further refined by Antuf'yev [61, 62], Mitskevich [63] and Brauer [64].

As previously noted, current practical needs require significant augmentation of heat transfer from finned surfaces. One of the ways of doing this is to design heat exchangers for operation at $Re > 10^5$. This has resulted in heightened interest in heat transfer and pressure drop for finned surfaces in fully developed turbulent flow.

It follows from this brief survey that the majority of studies of heat transfer and pressure drop for bundles of finned tubes were for the Reynolds number range from 10^3 to 2×10^4. Thus, they were performed over the ranges of developed mixed flow. Only a few studies were performed at $Re > 5 \times 10^4$, and these were not sufficient to provide detailed information about the relationships governing heat transfer from finned bundles over this range of Re. The available heat transfer correlations are for $Re < 5 \times 10^5$ [37, 38]; they are based on determinations of the average convective heat transfer coefficient and the nonuniformity in distribution of heat transfer over the fin surface is usually neglected entirely. This results in insufficient accuracy for the correlations. The available equations for calculating the nonuniformity of heat transfer over the surface of a finned tube have not been reliably verified under experimental conditions; indeed, the local characteristics of heat transfer from a finned tube are little explored in any sense.

To rectify, to some extent, the dearth of data mentioned in the previous paragraph, we have performed an extensive experimental study with bundles of tubes with spiral fins at Re from 10^4 to 10^6. The experiments were performed with a consistent technique and using the same test facility. The effect of fin geometry on heat transfer and hydraulic drag was investigated with 12 staggered bundles [67]. The effect of bundle configuration was studied with 9 such bundles.

The nonuniformity in distribution of heat transfer over the fin surface at high Re was investigated experimentally on tubes with spiral and circular fins, placed in the first and then in the fifth longitudinal row of a bundle, respectively [48–50].

In our studies, a great deal of attention was paid to investigating the physical pattern of heat transfer. We determined experimentally the local coefficients of heat transfer on the surface of a finned tube, for both the single tube and also with the tube placed in a bundle.

In these experiments, a large volume of experimental data was obtained. This made it possible to formulate a technique for the more exact determination of the fin effectiveness and of the convective heat transfer coefficients. This, in turn, allowed the development of more efficient forms of correlations of the experimental data. It was then possible to compare bundle effectiveness on the basis of various parameters to obtain working equations, for mixed and turbulent flow modes, which are convenient for heat exchanger design.

CHAPTER
TWO

CONDUCTION OF HEAT IN A FINNED TUBE

Heat transfer in bundles of finned tubes is more complicated than heat transfer in smooth-tube bundles. This is due to the complex nature of the fluid flow over the fin surface, the effect of the nonuniform temperature distributors over the fin, the effect of the complex geometry of the finned tube, and other factors. These factors interact closely with one another and complicate the solution of problems of heat transfer of finned surfaces.

In this chapter we present a brief survey of the principal analytic aspects of heat conduction in tubes with circular and spiral fins, which must be considered in investigating the heat transfer of bundles composed of such tubes.

2.1. FIN DESIGN

Usually, the fins are mounted on the outer surface of the tube. If heat is transferred from the outside of the tube to the fluid, then the fin temperature decreases in the radial direction. In the limit, when the tip of the fin attains the temperature of the fluid stream, the heat flux through the fin base attains its maximum value. An increase in the fin height beyond this point no longer increases the heat flux through the fin base. This limiting condition can be described by the expression

$$\frac{dQ}{dh} = 0 \qquad (2.1)$$

This equation was solved by Eckert and Drake [2] for a plane rectangular fin, assuring constant heat transfer coefficient and fin thermal conductivity and also as constant temperature across the fin. The solution has the form:

$$\frac{1}{\alpha} = \frac{\delta}{2\lambda} \qquad (2.2)$$

The left-hand side of the above equation represents the resistance to heat transfer between the fin and the fluid, whereas the right-hand side is resistance to thermal conduction in a plate corresponding to half the fin thickness. When these two resistances are equal, no benefit is gained by using these particular fins on the surface.

Since, however, the actual conditions differ from the assumptions underlying the derivation of Eq. (2.2), the use of fins is regarded as advantageous if

$$\frac{2\lambda}{\alpha\delta} > 5 \qquad (2.3)$$

However, actually the advisability of using fins is additionally controlled by factors such as the hydraulic drag of the finned surface and the weight and overall dimensions of the heat exchanger, and these factors must also be taken into account.

In designing heat exchangers it is sometimes very important to determine the conditions which give maximum heat transfer rate for a given exchanger weight. Solutions to this problem were obtained by Schmidt [1] and Eckert and Drake [2] who showed that the maximum heat dissipation through a rectangular plane fin of a given weight is obtained when:

$$\frac{2h}{\delta} = 1.419 \sqrt{\frac{2\lambda}{\alpha\delta}} \qquad (2.4)$$

The efficacy of the use of fins can be also determined in a different way. Let us assume that a surface has no fins. If it is assumed that the quantity of heat released by the area of the surface which is then covered by the fin base is equal to Q^1, whereas the amount of heat released by the finned surface is Q, then the ration Q/Q^1 will serve as a measure of the efficacy of heat dissipation by finned surfaces. For a plane rectangular fin defined by Eq. (2.4) Q/Q^1 is given by the expression:

$$\frac{Q}{Q'} = 0.889 \sqrt{\frac{2\lambda}{\alpha\delta}} \qquad (2.5)$$

It is seen that Eqs. (2.2)–(2.5) all contain the ratio $2\lambda/\alpha\delta$, which, as will be seen below, is also a very important parameter for circular fins. This parameter, written as $\alpha\delta/\lambda$, is termed the Biot number. It differs from the Nusselt number only by the fact that the thermal conductivity in the Biot number pertains to the fin wall metal.

The cross sectional area f of a rectangular plane fin, needed for a given heat dissipation rate Q can be obtained from the equation [2]:

$$f = \frac{2.109}{4L^3 d^2 \lambda} \left(\frac{Q}{\vartheta_1 - \vartheta_f} \right)^3 \tag{2.6}$$

where L is the fin length. Equation (2.6) shows that, in order to increase the amount of heat transferred through a finned surface it is best to increase the number of fins (i.e., L) and not the fin cross section. In addition, this equation makes it possible to clarify and compare the advantages of different fin materials. For example, it follows that the fin weight needed for a given Q is proportional to the ratio ρ/λ where ρ is the density of the fin material. Hence, in many practical cases, it is advantageous, from a weight point of view, to replace copper with aluminum and other light metals. Heavier materials such as carbon or stainless steel can be advantageous when their low cost per unit weight (or resistance to corrosion) offsets their high ρ/λ ratio. With fins of constant cross section, the upper part of the fin may not be utilized too efficiently. As the tip of the fin is approached, the temperature gradient falls off, which reduces the heat flux. The question arises of evaluating the optimal fin shape for which the heat transfer rate would be maximum for a given fin weight. It was pointed out as early as 1926 by Schmidt [1] that the heat flux through the cross section of such a fin should remain constant. This means that the heat flux lines should be parallel to the fin axis. Under these conditions, the temperature along the heat flux lines will vary linearly (Fig. 2.1). For any cross section of such a fin we can write

$$\vartheta - \vartheta_f = \frac{x}{h} (\vartheta_1 - \vartheta_f) \tag{2.7}$$

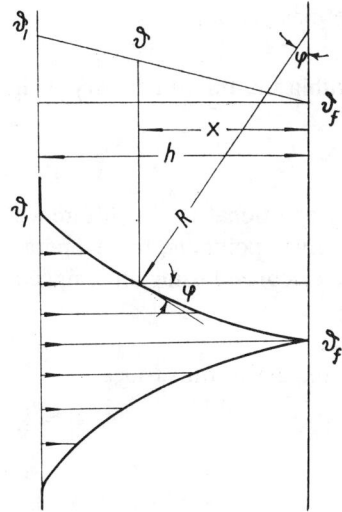

Figure 2.1 Fin with constant heat flux over its height.

Eckert and Drake [2] present a solution for this fin shape, according to which the equation defining angle φ as a function of distance x is written as:

$$\sin \varphi = \frac{\alpha(\vartheta_1 - \vartheta_f)}{qh} x \tag{2.8}$$

Here q is the heat flux density along the fin axis. The resultant fin profile is one bounded by arcs of radius R as shown in Fig. 2.1. Since $\sin \varphi = xR$, we have that $R = qh/\alpha(\vartheta_1 - \vartheta_f)$. Approximately the same conclusion was drawn by Makhin with his coworkers [3] in solving the differential equation for a trapezoidally-shaped fin.

Since the manufacture of fin shapes such as the form shown in Fig. 2.1 involves production difficulties, extensive use has been made of triangular cross section fins with a sharp tip and fins with truncated tip, i.e., trapezoidal fins, both of which approximate to the optimal shape.

The heat conduction through a straight circumferential fin will be considered in detail below. The cross section of such a fin varies radially and for this reason it is more difficult to solve the heat conduction equation than for the case of a plane rectangular fin with constant cross section.

2.2. TEMPERATURE GRADIENT IN A CIRCUMFERENTIAL FIN

A major problem in calculating heat transfer from a finned surface with circumferential fins is the estimation of the temperature distribution over the fin. This problem has been investigated in detail analytically by Schmidt [1], Il'in and Styrikovich [6], Gardner [7], Krischer and Kast [39], Kern [68] and Isachenko and coworkers [69].

Due to the complexity of the problem under study, analytical solutions are usually based on a number of simplifying assumptions [7, 68]:

1. The heat flux and temperature distribution within the fin do not vary with time (steady state).
2. The fin material is uniform and homogeneous.
3. There are no internal heat sources in the fin.
4. The heat flux on the fin surface is directly proportional to the difference between the temperature of the surface at the given point and the temperature of the ambient fluid (i.e., the flux can be calculated using an assigned heat transfer coefficient).
5. The thermal conductivity of the fin is constant.
6. The heat transfer coefficient is constant over the entire fin surface.
7. The fluid ambient temperature is constant.
8. The temperature of the fin base is constant.

CONDUCTION OF HEAT IN A FINNED TUBE 23

9. The temperature gradient in the direction perpendicular to the fin surface is negligible.
10. There is no thermal resistance at the point of contact between the tube and the fin.

For fins which are integral with the tube (where assumption 10 is valid) all these assumptions, apart from 6, 8 and 9, are likely to be closely valid.

Let us consider a tube with a circumferential fin of constant thickness δ (Fig. 2.2).

The ambient temperature (temperature of the fluid) ϑ_f is assumed to be constant. The fin temperature changes only in the radial direction, $\vartheta = f(r)$. The temperatures at the base and tip of the fin are respectively ϑ_1 and ϑ_2. The heat transfer coefficient for the fin surface is α. For an elementary ring with radii r and $d + dr$ under steady conditions we can write:

$$Q_r - Q_{r+dr} = dQ, \qquad (2.9)$$

$$Q_r = -2\pi\lambda\delta r \frac{d\vartheta}{dr}, \qquad (2.9a)$$

$$Q_{r+dr} = Q_r + \frac{dQ_r}{dr} dr = -$$

$$-2\pi\lambda\delta \left(\frac{d\vartheta}{dr} r + \frac{d^2\vartheta}{dr^2} r dr + \frac{d\vartheta}{dr} dr \right), \qquad (2.9b)$$

$$Q_r - Q_{r+dr} = dQ = 2\pi\lambda\delta \left(\frac{d^2\vartheta}{dr^2} r dr + \frac{d\vartheta}{dr} dr \right) \qquad (2.10)$$

But, on the other hand, the heat flux transmitted from the annular element of the fin to the surrounding fluid is:

$$dQ = 2\alpha (\vartheta - \vartheta_f) 2\pi r dr \qquad (2.11)$$

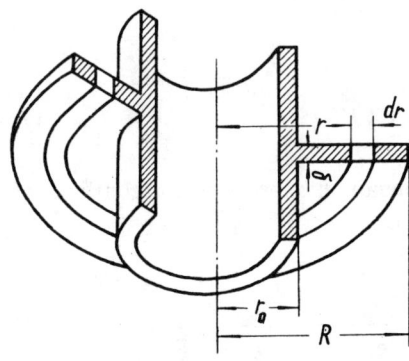

Figure 2.2 Schematic of a constant-thickness fin on a circular tube.

Equating the right-hand sides of Eqs. (2.10) and (2.11), and cancelling, we obtain:

$$\frac{d^2\vartheta}{dr^2} + \frac{1}{r}\frac{d\vartheta}{dr} - \frac{2\alpha}{\lambda\delta}\vartheta = 0 \qquad (2.12)$$

Designating

$$\sqrt{\frac{2\alpha}{\lambda\delta}} = \beta \text{ and } \beta r = z$$

we obtain:

$$\frac{d^2\vartheta}{dz^2} + \frac{1}{z}\frac{d\vartheta}{dz} - \vartheta = 0 \qquad (2.13)$$

This expression is a Bessel equation, which has a general solution in the form

$$\vartheta = C_1 I_0(z) + C_2 K_0(z) \qquad (2.14)$$

where $I_0(z) = I_0(\beta r)$ and $K_0(z) = K(\beta r)$ are zero-order Bessel functions of the imaginary argument.

We now perform calculations for the case of cooling of a fin in the presence of heat transfer at the tip. The boundary conditions at the tip are

$$-\lambda \left(\frac{d\vartheta}{dr}\right)_{r=R} = \alpha \vartheta_2 \qquad (2.15)$$

Additionally, the following equations hold at the fin base and its tip

$$\vartheta_1 = C_1 I_0(\beta r_0) + C_2 K_0(\beta r_0) \qquad (2.16)$$

and

$$\vartheta_2 = C_1 I_0(\beta R) + C_2 K_0(\beta R) \qquad (2.17)$$

Solving the equations simultaneously, we obtain an expression for constants of integration C_1 and C_2:

$$C_1 = \frac{\left[K_1(\beta R) - \frac{\alpha}{\lambda\beta} K_0(\beta R)\right] \vartheta_1}{I_0(\beta r_0) K_1(\beta R) + I_0(\beta R) K_0(\beta r_0) + \frac{\alpha}{\lambda\beta}[I_0(\beta R) K_0(\beta r_0) - I_0(\beta r_0) K_0(\beta R)]} \qquad (2.18)$$

CONDUCTION OF HEAT IN A FINNED TUBE

$$C_2 = \frac{\left[I_1(\beta R) + \frac{\alpha}{\lambda\beta} I_0(\beta R)\right]\vartheta_1}{I_0(\beta r_0) K_1(\beta R) + I_0(\beta R) K_0(\beta r_0) + \frac{\alpha}{\lambda\beta}[I_0(\beta R) K_0(\beta r_0) - I_0(\beta r_0) K_0(\beta R)]} \quad (2.19)$$

where I_1 and k_1 are first order Bessel functions. Using the expressions for the temperature gradient at the fin:

$$\vartheta = \vartheta_1 \frac{I_0(\beta r) K_1(\beta R) + I_1(\beta R) K_0(\beta r) + \frac{\alpha}{\lambda\beta}[I_0(\beta R) K_0(\beta r) - I_0(\beta r) K_0(\beta R)]}{I_1(\beta r_0) K_1(\beta R) + I_1(\beta R) K_0(\beta r_0) + \frac{\alpha}{\lambda\beta}[I_0(\beta R) K_0(\beta r_0) - I_0(\beta r_0) K_0(\beta R)]}$$

(2.20)

and also at its tip:

$$\vartheta_2 = \vartheta_1 \frac{I_0(\beta R) K_1(\beta R) + I_1(\beta R) K_0(\beta R)}{I_0(\beta r_0) K_1(\beta R) + I_1(\beta R) K_0(\beta r_0) + \frac{\alpha}{\lambda\beta}[I_0(\beta R) K_0(\beta r_0) - I_0(\beta r_0) K_0(\beta R)]}$$

(2.21)

If it is assumed that the surface of the tip does not participate in heat transfer, we can reduce Eqs. (2.20) and (2.21) to a simpler form. In this case the boundary condition at the tip is:

$$\left(\frac{d\vartheta}{dr}\right)_{r=R} = 0 \quad (2.22)$$

The constants of integration are defined as:

$$C_1 = -\frac{K_1(\beta R)\vartheta_1}{I_0(\beta r_0) K_1(\beta R) + I_1(\beta R) K_0(\beta r_0)} \quad (2.23)$$

$$C_2 = \frac{I_1(\beta R)\vartheta_1}{I_0(\beta r_0) K_1(\beta R) + I_1(\beta R) K_0(\beta r_0)} \quad (2.24)$$

The equation of the temperature gradient for this case is:

$$\vartheta = \vartheta_1 \frac{I_0(\beta r) K_1(\beta R) + I_1(\beta R) K_0(\beta r)}{I_0(\beta r_0) K_1(\beta R) + I_1(\beta R) K_0(\beta r_0)} \quad (2.25)$$

$$\vartheta_2 = \vartheta_1 \frac{I_0(\beta R) K_1(\beta R) + I_1(\beta R) K_0(\beta R)}{I_0(\beta r_0) K_1(\beta R) + I_1(\beta R) K_0(\beta r_0)} \quad (2.26)$$

Using the results of Il'in and Styrikovich [6] the heat transfer at the tip in this case can be incorporated with sufficient accuracy, by using a fin height increased by one half of a fin thickness.

In analyzing the expression for the temperature gradient of the fin, it is best to consider also another solution of the principal differential equation (2.12), obtained by Gardner [7]. This generalized equation, as shall be seen further down, makes it possible to determine not only the temperature gradient, but also the fin effectiveness. In solving this equation Gardner introduces still another assumption, in addition to the aforementioned 10, to the effect that the quantity of heat transmitted through the fin tip is negligible, as compared with that transmitted through the lateral surfaces.

Gardner suggests that Eq. (2.12) should be solved by equating it to the principal Bessel equation in the form

$$r^2 \frac{d^2 \vartheta}{dr^2} + r \frac{d\vartheta}{dr} + (r^2 - n^2) v = 0 \qquad (2.27)$$

where n is a constant. This yields the transformed equation:

$$r^2 \left(\frac{d^2 v}{dr^2}\right) + [(1 - 2m)r - 2ar] + [p^2 C_3^2 r^{2p} + a^2 r^2 + \qquad (2.28)$$
$$+ a(2m - 1)r + (m^2 - p^2 n^2)]\vartheta = 0.$$

where a, C_3, p, m and n are constants.

Equations (2.12) and (2.28) take the same form, if

$$f = C_4 r^{1 - 2pm} \qquad (2.29)$$

and

$$\frac{dF}{dr} = C_5 r^{2p(1-n) - 1} \qquad (2.30)$$

Here f is the cross sectional area of the fin, F is the fin surface area, c_4 and C_5 are positive constants.

Then the general solution of the equation can be obtained from Bessel functions using boundary conditions at the fin tip and base. At the tip

$$\left(\frac{d\vartheta}{dr}\right)_{r=R} = 0$$

(Gardner's additional assumption), and at the base — $(\vartheta)_{r=r_0} = \vartheta_1$.
At n equal to zero or to an integer, we obtain the equation of the temperature gradient

$$\vartheta = \vartheta_1 \left(\frac{u}{u_0}\right)^n \left[\frac{I_n(u) + \beta_1 K_n(u)}{I_n(u_0) + \beta_1 K_n(u_0)}\right] \qquad (2.31)$$

where

$$\beta_1 = -\frac{I_{n-1}(u_R)}{K_{n-1}(u_R)} \qquad (2.32)$$

and

$$u = -iC_3 r^p = r\sqrt{\frac{\alpha}{\lambda f}\frac{dF}{dr}} \qquad (2.33)$$

u_0 and u_R for the fin base and fin respectively are determined by substituting values of r, f and dF/dr, on the basis of conditions at the base and tip of the fin, $i = \sqrt{-1}$.

2.3. HEAT DISSIPATION FROM A CIRCUMFERENTIAL FIN

On the assumption that the entire heat flux passes through the fin base, we find the amount of heat transmitted by the fin to the flow:

$$Q = -\lambda f_0 \left(\frac{d\vartheta}{dr}\right)_{r=r_0} \qquad (2.34)$$

where

$$f_0 = 2\pi r_0 \delta, \quad \frac{d\vartheta}{dr} = [C_1 I_1(\beta r) - C_2 K_1(\beta r)]\beta$$

or

$$Q = \lambda f_0 \beta \vartheta_1 \psi \qquad (2.35)$$

where

$$\psi = \frac{I_1(\beta R) K_1(\beta r_0) - I_1(\beta r_0) K_1(\beta R) + \frac{\alpha}{\lambda\beta}[I_1(\beta r_0) K_0(\beta R) + I_0(\beta R) K_1(\beta r_0)]}{I_0(\beta r_0) K_1(\beta R) + I_1(\beta R) K_0(\beta r_0) + \frac{\alpha}{\lambda\beta}[I_0(\beta R) K_0(\beta r_0) - I_0(\beta r_0) K_0(\beta R)]} \qquad (2.36)$$

If no heat is transferred at the fin tip, we obtain

$$Q = \lambda f_0 \beta \vartheta_1 \psi' \qquad (2.37)$$

where

$$\psi' = \frac{I_1(\beta R) K_1(\beta r_0) - I_1(\beta r_0) K(\beta r)}{I_0(\beta r_0) K_1(\beta R) + I_1(\beta R) K_0(\beta r_0)} \qquad (2.38)$$

As previously noted, in practice, fins are not fabricated with a constant cross section, but are made as truncated triangles, i.e., trapezoids, and are thus close to the optimal shape discussed previously. No exact solution of the problem of heat transfer in a narrowing down, trapezoidally shaped, circumferential fin is available. Calculations can be performed with an approximate solution, obtained by subdividing the narrowing down fin into a series of constant-thickness rings [6]. The temperature gradient at the fin tip

$$\vartheta_2 = \vartheta_1 \frac{[K_1(\beta R) I_0(\beta R) + I_1(\beta R) K_0(\beta R)] - \dfrac{Q_T}{\lambda \beta F_T} [K_0(\beta r) I_1(\beta r_0) + I_0(\beta r_0) K_1(\beta r_0)]}{I_0(\beta r) K_1(\beta R) + I_1(\beta R) K_0(\beta r_0)} \qquad (2.39)$$

and factor ψ in Eq. (2.35) for the heat flux in this case

$$\psi = \vartheta_1 \frac{[I_0(\beta R) K_1(\beta r_0) - I_1(\beta r_0) K_1(\beta R)] + \dfrac{Q_T}{\lambda \beta F_T} [K_0(\beta r_0) I_1(\beta r_0) + I_0(\beta r_0) K_1(\beta r_0)]}{I_0(\beta r_0) K_1(\beta R) + I_1(\beta R) K_0(\beta r_0)} \qquad (2.40)$$

where Q_T is the quantity of heat released by the tip,

$$Q_T = -\lambda \beta [A I_0(\beta R) - B K_1(\beta R)] F_T \qquad (2.41)$$

and F_T is the surface of the fin tip.

Let us assume that there are no fins and that it is necessary to transmit by convection from the bare tube the same amount of heat Q as with fins. We introduce the conditional heat transfer coefficient α^* [39] and, assuming $\vartheta_F = 0$, we obtain:

$$Q = f_0 \alpha^* \vartheta_1 \qquad (2.42)$$

Here, α^* is the heat transfer coefficient necessary for the transmission of an amount of heat equal to that for a given finned tube, but from a base tube whose diameter is equal to the fin base diameter. It follows from Eqs. (2.9a) and (2.42) that:

$$\alpha^* = -\frac{\lambda}{\vartheta_1} \left(\frac{d\vartheta}{dr}\right)_{r=r_0} \qquad (2.43)$$

Differentiating according to (2.20) and (2.35), we obtain:

$$\alpha^* = \lambda \beta \psi \qquad (2.44)$$

Figure 2.3 shows a plot of $\alpha^*/\lambda\beta$ vs the nondimensional fin height βh for different nondimensional radii of the fin-supporting tube βr_0. The curves pertain to the case when $(\alpha/\lambda\beta) r = R = 0$ (i.e., when there is no transfer of heat at the tip) and to two other cases, when $(\alpha/\lambda\beta) r = R \neq 0$. These three parameters $\alpha^*/\lambda\beta$, βh and βr_0 can (according to Krischer with Kast [39]) be expressed as:

$$\frac{\alpha^*}{\lambda\beta} = \frac{\alpha^*}{\alpha}\sqrt{\frac{\alpha\delta}{2\lambda}}, \quad \beta h = \frac{2h}{\delta}\sqrt{\frac{\alpha\delta}{2\lambda}}, \quad \beta r_0 = \frac{2r_0}{\delta}\sqrt{\frac{\alpha\delta}{2\lambda}} \qquad (2.45)$$

whence it is seen that all of them contain one common nondimensional parameter

$$\sqrt{\frac{\alpha\delta}{2\lambda}}$$

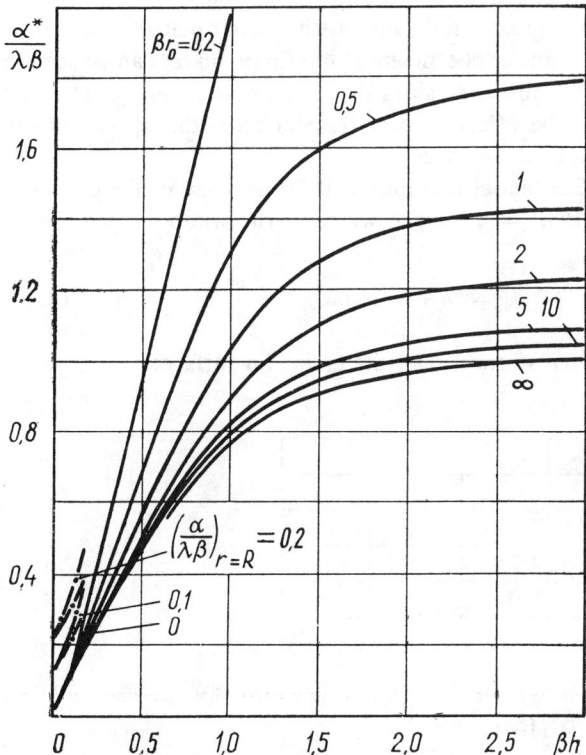

Figure 2.3 Plot of $\alpha^*/\lambda\beta = - = f(\beta h)$ at a different βr_0 and $(\alpha/\lambda\beta) r = R$.

30 HEAT TRANSFER OF FINNED TUBE BUNDLES IN CROSSFLOW

This is the great advantage of these expressions. Thus, the ratio α^*/α, indicating the gain in heat flux transmitted by the finned tube as compared with one without fins (Fig. 2.3), is related directly only to parameter

$$\sqrt{\frac{\overline{\alpha\delta}}{2\lambda}}$$

It follows from Fig. 2.3 that the nondimensional heat transfer coefficient $\alpha^*/\lambda\beta$ first rises steeply with increasing βh, and then gradually approaches the limiting value

$$\lim\left(\frac{\alpha^*}{\lambda\beta}\right) = \frac{K_1(\beta r_0)}{K_0(\beta r_0)} \qquad (2.46)$$

At $\beta h \simeq 2$ the value of $\alpha^*/\lambda\beta$ is already quite close to its limiting value. This indicates that values of βh in excess of 2 should not be used for practical purposes. It is additionally seen from Fig. 2.3 that the value of $\alpha^*/\lambda\beta$ increases with reduction in radius r_0 of the fin-bearing tube. The value of α^* for tubes with identical fins is always greater for a tube with smaller diameter d.

The effect of the heat transfer coefficient at the fin tip on α^* can be seen in Fig. 2.4. In the case of low fins the values of $(\alpha/\lambda\beta)\ r = R$ are greater and should not be neglected. The effect of heat transfer from the tip is actually perceptible over the range $0 < \beta h \leq 2$.

It is seen from Fig. 2.3 that at low βh (< 0.2) the slope of the curves is approximately $d\,(\alpha^*/\lambda\beta)/d(\beta h) \simeq 1$. Hence we can write without much error:

$$\alpha^*/\lambda\beta = \beta h + (\alpha/\lambda\beta)_{r=R} \qquad (2.47)$$

and, accordingly, if $\alpha_R = \alpha = $ const over the entire fin surface:

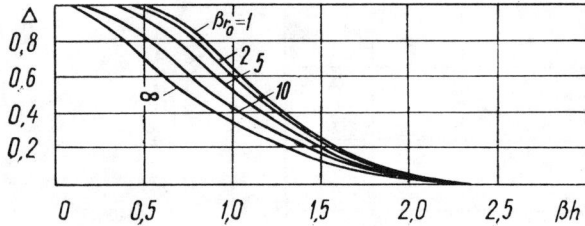

Figure 2.4 Determination of the effect of the fin tip heat transfer coefficient α_R on the conditional heat transfer coefficient α^*, $\Delta = f(\beta h)$.

$$\Delta = \frac{(\alpha^*/\lambda\beta)\alpha_{R\neq 0} - (\alpha^*/\lambda\beta)\alpha_{R=0}}{(\alpha/\lambda\beta)_{r=R}}$$

$$\alpha^* = \lambda\beta^2\, h + \alpha = \lambda\, \frac{2\beta}{\delta\lambda}\, h + \alpha = \alpha\left(1 + \frac{2h}{\delta}\right) \qquad (2.48)$$

Equation (2.48) above can also be written thus:

$$\alpha^* = \alpha\, \frac{h + \delta/2}{\delta/2} = \alpha\, \frac{2h'}{\delta} \qquad (2.49)$$

where h^1 is the height of the fin, increased by one half of its thickness. The above provides an opportunity to estimate relative simply the heat transfer from the fin tip.

If heat transfer at the tip is neglected, then we have for $\beta h < 0.2$ the expression

$$(\alpha^*)_{\alpha R = 0} = \alpha\, \frac{2h}{\delta} \qquad (2.50)$$

It is seen that the difference in conditional heat transfer coefficient α^* predicted Eqs. (2.48) and (2.50) respectively is 50% for $2h/\delta = 2$, whereas at $2h/\delta = 10$ it is only 10%.

It follows from the above that, whereas at very low βh it is possible to assume without much error that $\alpha^* \simeq \alpha$, at higher values of βh (and already for $\beta h > 2$), $\alpha^*/\lambda\beta$, i.e.,

$$\alpha^* \simeq \lambda\beta = \sqrt{2\alpha\lambda/\delta} \simeq \sqrt{\alpha}$$

It can be concluded from this that there is a limit to the increase of conditional heat transfer coefficient α^* of each fin with increasing height, is when the quantity of heat transmitted by the fin no longer increases with height. This limit occurs at $\alpha^* < \alpha$.

2.4. HEAT CONDUCTION OF SPIRAL FINS

The problem of heat conduction of spiral fins is addressed using the same assumptions as those used in deriving the differential equation for circumferential fins.

The general differential equation for spiral fins with trapezoidal cross section is obtained by writing a heat balance over a spiral-shaped element of height dr between $AA'D'D$ and $CC'B'B$ (Fig. 2.5). It is written as [70]:

$$\frac{d^2\vartheta}{dr^2} + \frac{1}{z(r)}\, \frac{d\vartheta}{dr}\, \frac{dz(r)}{dr} + \frac{r}{s^{12} + r^2}\, \frac{d\vartheta}{dr} - \frac{\alpha}{\lambda z(r)}\, \vartheta = 0 \qquad (2.51)$$

This equation has no solution with the exception of the particular case [70] of a

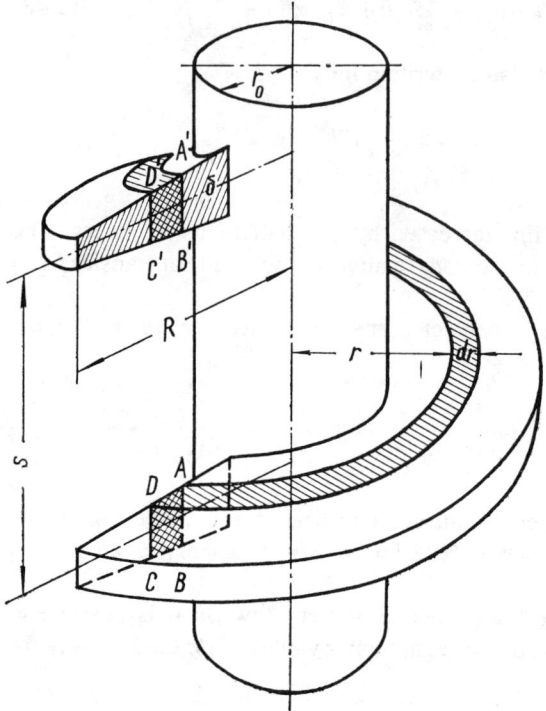

Figure 2.5 Continuous spiral fin with a trapezoidal cross section.

rectangular fin cross section, i.e., heightwise constant thickness (Fig. 2.6):

$$\frac{dz(r)}{dr} = 0$$

The principal differential equation for this condition is written as:

$$\frac{d^2\vartheta}{dr^2} + \frac{r}{s'^2 + r^2} \frac{d\vartheta}{dr} - \beta\vartheta = 0 \qquad (2.52)$$

where

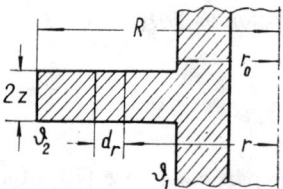

Figure 2.6 Fin with rectangular cross section.

$$\frac{r}{s'^2+r^2}\,\frac{d\vartheta}{dr} -$$

is the curvature of the spiral.

The general solution of this equation is

$$\vartheta = \frac{1}{(s'^2+r^2)^{\frac{1}{4}}} [C_1 e^{\beta r} + C_2 e^{-\beta r}] \tag{2.53}$$

C_1 and C_2 are constants of integration, obtained from the boundary conditions

$$r = r_0, \quad \vartheta = \vartheta_1$$
$$r = R, \quad \vartheta = \vartheta_2$$

and are written as:

$$C_1 = \frac{\vartheta_1 (s'^2+r_0^2)^{\frac{1}{4}} e^{-\beta R} + \vartheta_2 (s'^2+R^2)^{\frac{1}{4}} e^{-\beta r_0}}{e^{\beta(R-r_0)} - e^{-\beta(R-r_0)}} \tag{2.54}$$

$$C_2 = \frac{\vartheta_1 (s'^2+r_0^2)^{\frac{1}{4}} e^{\beta R} - \vartheta_2 (s'^2+R^2)^{\frac{1}{4}} e^{\beta r_0}}{e^{\beta(R-r_0)} - e^{-\beta(R-r_0)}} \tag{2.55}$$

Using the expressions for C_1 and C_2, we can write the equation for the temperature gradient over the fin in the final form:

$$\vartheta = (s'^2+r^2)^{\frac{1}{4}} \left[\vartheta_1 (s'^2+r_0^2) \frac{\sinh \beta (R-r)}{\sinh \beta (R-r_0)} - \right.$$
$$\left. - \vartheta_2 (s'^2+R^2) \frac{\sinh \beta (R-r)}{\sinh \beta (R-r_0)} \right] \tag{2.56}$$

The amount of heat transmitted from a spiral fin is determined, as for a circumferential fin, from the condition that the entire heat flux passes through the fin-base area (see Eq. (2.34)). In this case the area of the base is calculated from the expression

$$f_0 = 2z(r) \sqrt{s^2 + (2\pi r_0)^2} \tag{2.57}$$

The equation for the quantity of transmitted heat is expressed in the final form [70]:

$$Q = 4\pi\lambda\delta \, (s'^2 - r_0^2) \, \beta \times$$
$$\times \left\{ \vartheta_1 \left(\cot h \left[\beta \left(R - r_0 \right) \right] + \vartheta_2 \left(\frac{s'^2 + R^2}{s'^2 + r_0^2} \right)^{\frac{1}{4}} \cos ech \left[\beta \left(R - r_0 \right) \right] \right\} \quad (2.58)$$

2.5. FIN EFFECTIVENESS

In normal heat transfer, the heat flux transmitted from one fluid through a wall to another fluid, is often directly proportional to the area of the wall and the temperature difference between the fluids. If the heat transfer area on one side of the wall is increased by thin metal fins, it might be postulated that the heat flux referred to unit surface of the wall carrying the fins will increase in direct proportion to the heat transmission surface area. However, due to the existence of a temperature gradient along the fin, the effective temperature difference decreases, for which reason the overall increase in the heat flux will be smaller than expected.

To facilitate the calculation of heat transfer of a finned surface one uses the concept of fin effectiveness, which indicates the extent to which heat transfer is augmented by putting fins on a given surface.

The fin effectiveness is defined as the ratio of the amount of heat transmitted by the finned surface to that which could be transmitted if the fin had infinite thermal conductivity.

If the amount of heat transmitted by the fin surface to the fluid flowing over it represented in the form

$$Q = \int_0^{F_p} \alpha \, (\vartheta - \vartheta_f) \, dF_{\text{fin}} \quad (2.59)$$

where α is the heat transfer coefficient of the fin, then assuming the thermal conductivity of the fin to be infinite, the employing assumptions 6 and 7 (section 2.2 above), we find that the temperature of the fin surface approaches ϑ_1, the temperature of its base. Then

$$Q = (\vartheta_1 - \vartheta_f) \int_0^{F_p} \alpha \, dF_{\text{fin}} \quad (2.60)$$

Equations (2.59) and (2.60) yield an expression for the fin effectiveness:

$$E = \frac{\int_0^{F_p} \alpha(\vartheta - \vartheta_f) \, dF_{\text{fin}}}{(\vartheta_1 - \vartheta_f) \int_0^{F_p} \alpha \, dF_{\text{fin}}} \tag{2.61}$$

According to condition 6 (α is constant), we write:

$$E = \frac{\int_0^{F_p} (\vartheta - \vartheta_f) \, dF_{\text{fin}}}{(\vartheta_1 - \vartheta_f) F_{\text{fin}}} \tag{2.62}$$

or

$$E = \frac{\overline{\vartheta} - \vartheta_f}{\vartheta_1 - \vartheta_f} \tag{2.63}$$

whence it is seen that the fin effectiveness can be represented by the ratio of the mean temperature difference between the finned surface and the fluid to the temperature difference between the surface supporting the fins and the fluid. Equations (2.59) and (2.63) then yield

$$Q = \alpha(\vartheta_1 - \vartheta_f) E F_{\text{fin}} \tag{2.64}$$

Equations (2.42) and (2.64) define the same quantities of heat Q. Equating their right-hand sides, we obtain

$$E = \frac{\alpha^*}{\alpha} \frac{f_0}{F_{\text{fin}}} \tag{2.65}$$

or

$$\frac{\alpha^*}{\alpha} = \frac{F_{\text{fin}}}{f_0} E \tag{2.66}$$

On the basis of Eq. (2.31) Gardner [7] suggested the following expression for the fin effectiveness:

$$E = \frac{2(1-n)}{u_0 [1 - (u_R/u_0)^{2(1-n)}]} \left[\frac{I_{n-1}(u_0) - \beta_1 K_{n-1}(u_0)}{I_n(u_0) + \beta_1 K_n(u_0)} \right] \tag{2.67}$$

The exponent of r for a given fin cross section can be determined from Eq. (2.29).

At $n = 0$, equations (2.31) and (2.32) and also the expression

$$dF = C_0 u^{1-2n} du \qquad (2.68)$$

(obtained from Eqs. (2.30) and (2.33)) can be used for transforming Eq. (2.67) to a form suitable for determining the effectiveness of a circumferential fin of constant thickness, which is:

$$E = \frac{2}{\beta r_0 \left(1 - \dfrac{R}{r_0}\right)^2} \left[\frac{I_1(\beta r_0) - \beta_1 K_1(\beta r_0)}{I_0(\beta r_0) + \beta_1 K_0(\beta r_0)}\right] \qquad (2.69)$$

where

$$\beta_1 = \frac{I_1(\beta R)}{K_1(\beta R)}$$

Proceeding similarly, expressions for the fin effectiveness at different values of n were obtained for transverse circumferential fins with different cross sectional shapes. As previously noted, no correction was made in these solutions for heat transfer from the fin tip, which can result in significant error in the case of relatively low fins. We recall that heat transfer from the tip can be incorporated relatively simply by conditionally increasing the fin height by one half of its thickness.

Figure 2.7 shows a graph of fin effectiveness, calculated from Eq. (2.69) for circumferential fins of constant thickness as a function of βh.

Shneider [72] compared the effectiveness of circumferential fins with rectangular and converging (hyperbolic) cross section and found that for given height of both fins and for the same cross sectional areas f_0 the effectiveness of the hyperbolic fin is higher, and that this difference increases with βh.

It is easily seen that the fin effectiveness $E = f(\beta h) = f(h/\delta \sqrt{2Bi})$ tends to its maximum value, equal to 1, at $h/\delta \sqrt{2Bi} \to 0$ [69]. For specified fin dimensions the latter is possible at $\lambda \to \infty$.

The effectiveness of a spiral fin is defined by Kawashima and Katayama [70] thus:

$$E = \frac{\displaystyle\int_{r_0}^{R} 2\alpha\vartheta \sqrt{s^2 + (2\pi r)^2}\, dr}{2\alpha\vartheta_1 \displaystyle\int_{r_0}^{R} \sqrt{s^2 + (2\pi r)^2}\, dr} \qquad (2.70)$$

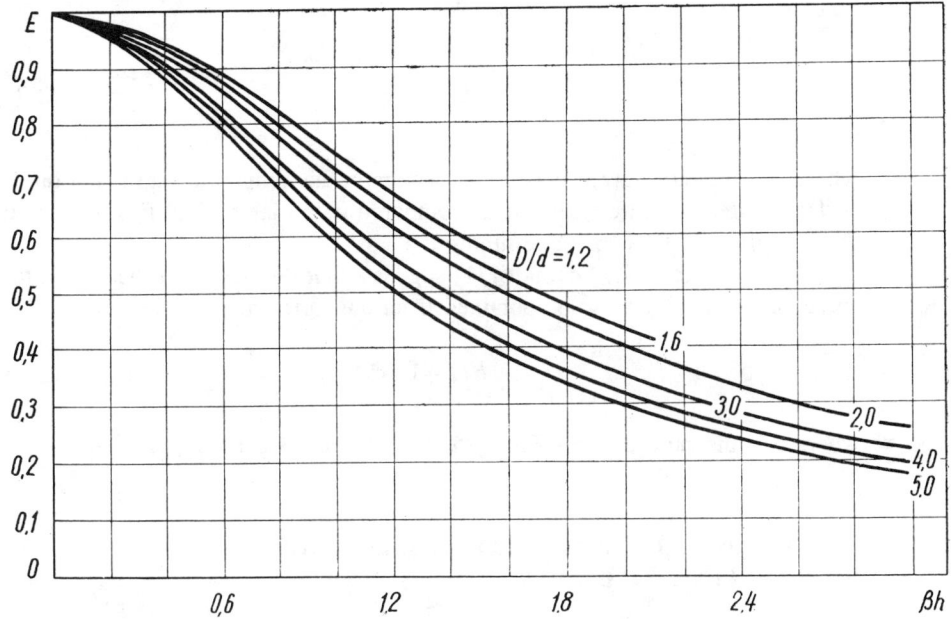

Figure 2.7 Effectiveness of transverse circumferential fins with rectangular cross section.

or

$$E = \frac{\lambda\delta\,(s'^2+r_0^2)^{\frac{1}{2}}\,\beta\left\{\vartheta_1\,\text{arc cot h}\,[\beta\,(R-r_0)] + \vartheta_2\left(\frac{s'^2+R^2}{s'^2+r_0^2}\right)^{\frac{1}{4}}\text{cosech}\,[\beta\,(R-r_0)]\right\}}{\alpha\,\vartheta_1\left[R\,(s'^2+R^2)^{\frac{1}{2}} - r_0\,(s'^2+r_0^2) + s'^2\log\left\{\frac{R+\sqrt{s'^2+R^2}}{r_0+\sqrt{s'^2+r_0^2}}\right\}\right]} \quad (2.71)$$

Here it was assumed $\vartheta_f = 0$.

A graphic expression for the effectiveness of spiral fins is not available.

A coefficient for the effectiveness of linear straight fins was determined by the function

$$E = \frac{\tanh x}{x} \quad (2.72)$$

where $x = \beta h$. Due to its convenient and simple form this expression is frequently used in the literature for calculating the effectiveness of circular fins.

The VDI Warmeatlas handbook [21] discusses ways of using the above expression. It is suggested that the equivalent height of the circumferential fin be expressed as

$$h' = \frac{d}{2} \varphi \qquad (2.73)$$

where φ is a factor correcting for the change in the temperature gradient on the surface of the circumferential fin as compared with longitudinal fins. (A nomogram is given in [21] for determining it.)

On the other hand, Schmidt [71] suggests that the height of the circumferential fin for this purpose be obtained from the formula:

$$h' = h\,(1 + 0.35 \ln \rho) \qquad (2.74)$$

where ρ is the ratio of the radii of the fin (R) and its carrying tube (r_0).

2.6. THERMAL CONDUCTIVITY OF THE TUBE SUPPORTING THE FINS

The general equation of heat conduction in a finned tube (2.13) pertains to the so-called one-dimensional theory of heat transfer, where particular attention is given to heat transfer in the fin proper and no consideration is given to features of heat transfer in the tube segment between the fins and beneath them.

Neglect of the heat conduction characteristics of the tube can, under certain design and performance conditions, result in significant errors. Hence, when necessary, one must use two-dimensional methods for heat transfer calculations. The basis of the different two-dimensional methods consists of reducing the general heat conduction equation to an equation in finite differences, which is then solved by one of the available numerical methods. An extensively used method for solving such problems is that of "elementary balances" due to Vanichev [73], according to which the heat transmitting body is subdivided into a number of simple geometric elements with linear temperature variations within each. The Fourier and Newton laws are applied to each such element and heat balances are written accordingly, which then yields n algebraic equations with n unknown temperatures. The resulting set of equations can be solved by one of the available numerical methods.

All the numerical methods have the disadvantage of being cumbersome and time consuming. As a result, in a number of studies, techniques were developed to account approximately for the two-dimensional nature of heat transfer.

We now consider in detail the solution of the two-dimensional problem for a tube with circular transverse fins [39].

Transfer of heat from liquid f_1, flowing over the finned tube on the outside, to liquid f_2, flowing inside the tube, is described by the equation

$$Q = kF(\vartheta_{f_1} - \vartheta_{f_2}) \tag{2.75}$$

where k is the heat transfer coefficient referred to the total outer surface of the finned tube.

The thermal resistance of the tube under study can be represented as

$$\frac{1}{kF} = \frac{1}{\alpha_0 F_{sm} + \alpha E F_p} + \frac{\delta'}{\lambda_w F_{int}} + \frac{1}{\alpha_{int} F_{int}} \tag{2.76}$$

where f_{sm} is the surface of the bare tube between the fins, α_0 is the coefficient of heat transfer corresponding to this surface, δ^1 is the thickness of the tube wall, and α_{int} is the coefficient of heat transfer corresponding to the inner tube surface F_{int}.

This equation does not take account of the logarithmic nature of the temperature field within the tube wall (which is significant in thin-walled tubes made of highly conductive metal), and also the axially periodic temperature field, corresponding in frequency to the fin spacing.

The finned tube is frequently assumed to behave as a bare tube, with isotherms in the base tube wall which are parallel to its axis. Such simplification is entirely valid only in the case of a thick-walled tube with closely spaced fins, since the aforementioned periodic temperature field equalizes rapidly within the wall thickness. The question arises, however, of the error in determining the heat flux from a finned tube with a thinner base tube wall and a wider fin pitch; this error arrives from the assumption of parallel isotherms. To evaluate this error, we shall consider the periodic temperature field within the tube wall.

Under steady conductive conditions, i.e., when the temperature field does not vary with time, the temperature field ($\partial\vartheta/\partial\phi = 0$) in a cylindrical body, when symmetrical to its axis, is described by the differential Laplace equation:

$$\nabla^2 \vartheta = \frac{\partial^2 \vartheta}{\partial r^2} + \frac{1}{r}\frac{\partial \vartheta}{\partial r} + \frac{\partial^2 \vartheta}{\partial z^2} = 0 \tag{2.77}$$

where r is the radial and z is the axial coordinate. Using the notation shown in Fig. 2.8, and taking n to be a free parameter, we can write, using constants A_0, A_1, A_2 and B_0, B_1 and B_2, the general solution of this equation [39]:

$$\vartheta_{(r,\,z)} = A_0 \left[I_0(nr) + B_0 K_0(nr)\right] \sin(nz) + A_1 \left[I_0(nr) + B_1 K_0(nr)\right] \cos(nz) +$$
$$+ A_2 \left[\ln(r/r_{int}) + B_2\right] \tag{2.78}$$

Taking the center of the fin (as seen in Fig. 2.8), as the origin for the axial z coordinate, we find that $A_0 = 0$ (since $\vartheta(z) = \vartheta(-z)$). The remaining constants (A_1, B_1, A_2, B_2) may be determined from boundary conditions.

40 HEAT TRANSFER OF FINNED TUBE BUNDLES IN CROSSFLOW

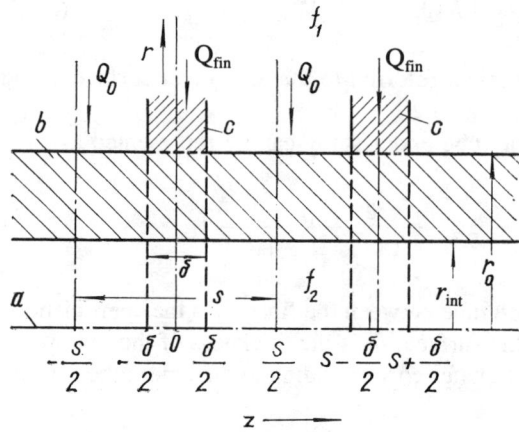

Figure 2.8 Schematic of finned tube. *a)* Tube axis; *b)* tube wall; *c)* fin.

The first and second boundary conditions arise from the fact that the temperature fields at $z = 0$ and at the midpoint between the fins are symmetrical, thus:

$$(\partial \vartheta / \partial z)_{z=0} = 0 \qquad (2.79)$$

$$(\partial \vartheta / \partial z)_{z=\frac{s}{2}} = 0 \qquad (2.80)$$

At the inner surface of the tube ($r = r_{int}$) the heat flux is transmitted to the internal fluid f_2, hence (3rd boundary condition):

$$-\lambda \, (\partial \vartheta / \partial r)_{r=r_{int}} = -\alpha_{int} [\, \vartheta \, (r_{int}, \, z) - \vartheta_{f_2}] \qquad (2.81)$$

This expression is also suitable for the outer surface of the tube ($r = r_0$) (though with a different heat transfer coefficient). The heat flux to the tube through the fin is as follows:

$$Q_{fin} = 2\pi r_0 \, \delta \alpha^* \, [\vartheta_{f_1} - \vartheta_{(r_0, z)}] \qquad (2.82)$$

Thus, on the tube segment between the fins

$$Q_0 = 2\pi r_0 (s - \delta) \, \alpha_0 \, [\vartheta_{f_1} - \vartheta_{(r_0, z)}] \qquad (2.83)$$

Hence we can write the following equation for heat transfer for tube segment

$$\tfrac{1}{2} \delta < z < \tfrac{1}{2} \delta$$

(4th boundary condition):

$$-\lambda \, (\partial \vartheta / \partial r)_{r=r_0} = -\alpha^* \, [\vartheta_{f_1} - \vartheta_{(r_0, z)}] \qquad (2.84)$$

and for segment

$$\frac{1}{2}\delta < z < s - \frac{1}{2}\delta$$

respectively,

$$-\lambda \, (\partial \vartheta / \partial r)_{r=r_0} = -\alpha_0 \, [\vartheta_{f_1} - \vartheta_{(r_0, z)}] \qquad (2.85)$$

In order to represent the last condition, a solution must be found for two different heat transfer coefficients (α^* and α_0).

Actually Eqs. (2.84) and (2.85) are two equations for which it is impossible to obtain a single solution. However, if it is assumed that the heat flux q_0 at the fin base and the heat flux q_0 between the fins is constant, then the heat flux at the outer tube surface can be determined by writing and solving a Fourier series. In this case

$$q_{(r_0, z)} = -\lambda \left(\frac{\partial \vartheta}{\partial r}\right)_{r=r_0} = c_0 + \sum_{j=1}^{\infty} c_{nf} \cos(n_j z) \qquad (2.86)$$

where $j = 1, 2, 3, \ldots$ is the ordinal number, and c_0 and c_{nf} are coefficients, obtainable from the expressions

$$c_0 = \frac{1}{s} \int_0^s q_{(r_0, z)} \, dz, \qquad c_{nf} = \frac{2}{s} \int_0^s q_{(r_0, z)} \cos(n_j z) \, dz \qquad (2.87)$$

Equations (2.79)–(2.87) allow one to determine the constants of Eq. (2.78). Boundary condition (2.79) is satisfied already at $A_0 = 0$. Condition (2.80) yields for

$$z = \frac{1}{2} s$$

that $\sin(n \cdot s/2) = 0$, and accordingly, $n \cdot s/2 = 0, \pi, 2\pi \ldots$, consequently,

$$n_j = j \, 2\pi/s \qquad (2.88)$$

For simplicity's sake n shall henceforth be written without a subscript. In addition, we assume that $\vartheta_{f_2} = 0$. Then condition (2.81) is written as:

$$-\lambda \, (\partial \vartheta / \partial r)_{r=r_{int}} = -\alpha_{int} \, \vartheta_{(r_{int}, z)} \quad (2.89)$$

Substituting the above into Eq. (2.78), we write:

$$-\lambda \left\{ A_1 n \cos (nz) \left[I_0 (nr_{int}) - B_1 K_1 (nr_{int}) \right] + A_2 \frac{1}{r_{int}} \right\} =$$
$$= -\alpha_{int} \left\{ A_1 n \cos (nz) \left[I_0 (nr_{int}) + B_1 K_0 (nr_{int}) \right] + A_2 B_2 \right\} \quad (2.90)$$

Equating the coefficients of terms, dependent and independent of z, we find:

$$B_2 = \lambda / \alpha_{int} \, r_{int} \quad (2.91)$$

$$B_1 = \frac{I_0(nr_{int}) + (n\lambda / \alpha_{int}) I_1 (nr_{int})}{K_0 (nr_{int}) + (n\lambda / \alpha_{int}) K_1 (nr_{int})} \quad (2.92)$$

According to boundary conditions (2.84) and (2.85), we find the heat flux density $q \, (r_0, z) = -\lambda \, (\partial \vartheta / \partial r)_{r=r_{int}}$ at the outer tube surface ($r = r_0$) at segment

$$-\frac{1}{2} \delta < z < \frac{1}{2} \delta$$

$$q_{(r_0, z)} = -\alpha^* \{ \vartheta_{f_1} - A_1 \cos (nz) \left[I_0 (nr_0) + B_1 K_0 (nr_0) \right] -$$
$$- A_2 \left[\ln \left(\frac{r_0}{r_{int}} \right) + B_2 \right] \} \quad (2.93)$$

and at segment

$$\frac{1}{2} \delta < z < s - \frac{1}{2} \delta$$

$$q_{(r_0, z)} = -\alpha_0 \{ \vartheta_f - A_1 \cos (nz) \left[I_0 (nr_0) + B_1 K_0 (nr_0) \right] -$$
$$- A_2 \left[\ln \left(\frac{r_0}{r_{int}} \right) + B_2 \right] \}. \quad (2.94)$$

Substituting Eqs. (2.93) and (2.94) into (2.87), and integrating the equation thus obtained for both segments, we obtain Fourier series (2.86) in the form:

CONDUCTION OF HEAT IN A FINNED TUBE

$$q_{(r_0,z)} = -\left\{\left[\alpha^* \frac{\delta}{s} + \alpha_0 \left(1 - \frac{\delta}{s}\right)\right]\left[\vartheta_{f_1} - A_2\left(\ln\left(\frac{r_0}{r_{int}}\right) + B_2\right)\right] - \right.$$

$$- \sum_{j=1}^{\infty} A_1 (\alpha^* - \alpha_0) \frac{\sin(n\delta/2)}{n \cdot s/2} [I_0(nr_0) + B_2 K_0(nr_0)] \bigg\} -$$

$$- \sum_{j=1}^{\infty} \cos(nz) \left\{\frac{4}{s}(\alpha^* - \alpha_0) \frac{\sin(n\delta/2)}{n \cdot s/2}\left[\vartheta_{f_1} - A_2\left(\ln\left(\frac{r_0}{r_{int}}\right) + B_2\right)\right] - \right.$$

$$\left. - A_1\left[\alpha^* \frac{\delta}{s} + \alpha_0\left(1 - \frac{\delta}{s}\right) + \frac{\sin n\delta}{ns}(\alpha^* - \alpha_0)\right][I_0(nr_0) + B_2 K_0(nr_0)]\right\} \quad (2.95)$$

According to Eqs. (2.84) and (2.85), this expression should be equal to:

$$-\lambda \left(\frac{\partial \vartheta}{\partial r}\right)_{r=r_0} = -$$

$$-\lambda \left\{\sum A_1 n \cos(nz) [I_1(nr_0) - B_1 K_1(nr_0)] + A_2 \frac{1}{r_0}\right\} \quad (2.96)$$

From the last equations we can determine constants A_1 and A_2. For simplicity we employ the notation

$$\bar{\alpha} = \alpha^* \frac{\delta}{s} + \alpha_0 \left(1 - \frac{\delta}{s}\right) \quad (2.97)$$

where $\bar{\alpha}$ is the equivalent coefficient of heat transfer of a finned tube, obtained by averaging the respective coefficients α^* and α_0 of surfaces participating in heat transfer.

We can then write constants A_1 and A_2 as:

$$A_{1j} = \frac{2 \frac{\alpha^* - \alpha_0}{\bar{\alpha}} \frac{\sin(n\delta/2)}{n \cdot s/2} \vartheta_{f_1}}{r_0 \bar{\alpha}[(1/r_0\bar{\alpha}) + (1/\lambda)\ln(r_0/r_{int}) + (1/r_{int}\,\alpha_{int})]} \cdot \frac{1}{M_j}$$

$$\cdot \frac{1}{I_0(nr_0) + B_1 K_0(nr_0)} \quad (2.98)$$

$$A_{2j} = \frac{\vartheta_{f_1}}{\lambda} \left\{\frac{1}{1/r_0\bar{\alpha} + (1/\lambda)\ln(r_0/r_{int}) + (1/r_{int}\,\alpha_{int})} - \right.$$

$$\left. - \frac{2\left(\frac{\alpha^* - \alpha_0}{\bar{\alpha}}\right)^2 \left[\frac{\sin(n\delta/2)}{ns/2}\right]^2}{M_j r_0 \bar{\alpha}\,[(1/r_0\,\bar{\alpha}) + (1/\lambda)\ln(r_0/r_{int}) + (1/r_{int}\,\alpha_{int})]^2}\right\} \quad (2.99)$$

where n is obtained from the expression for n_j (2.88), and B_1 from (2.92)

$$M_j = 1 + \frac{n\lambda}{\bar{\alpha}} \frac{I_1(nr_0) - B_1 K_1(nr_0)}{I_0(nr_0) + B_1 K_0(nr_0)} \tag{2.100}$$

We introduce the notation:

$$nr_0 = 2\pi j r_0/s = \rho_0, \qquad nr_{\text{int}} = 2\pi j r_{\text{int}}/s = \rho_{\text{int}},$$
$$nr = 2\pi j r/s = \rho,$$
$$n\delta/2 = \pi j \delta/s = \bar{\delta}, \qquad ns/2 = \pi j,$$
$$nz = 2\pi j z/s = \xi, \qquad n\lambda/\alpha_{\text{int}} = 2\pi j \lambda/s \alpha_{\text{int}} = \mu \tag{2.101}$$

according to which Eqs. (2.100) and (2.92) become

$$M_j = 1 - \frac{2\pi j \lambda}{s\bar{\alpha}} \frac{I_1(\rho_0) - B_1 K_1(\rho_0)}{I_0(\rho_0) + B_1 K_0(\rho_0)} \tag{2.102}$$

$$B_1 = \frac{-I_0(\rho_{\text{int}}) + \mu I_1(\rho_{\text{int}})}{K_0(\rho_{\text{int}}) + \mu K_1(\rho_{\text{int}})} \tag{2.103}$$

Substitution of Eqs. (2.98) and (2.99) into (2.78), yields the general equation of the temperature field:

$$\vartheta_{(r,z)} = \sum_{j=1}^{\infty} A_{2j} \lambda \left[\frac{1}{\lambda} \ln\left(\frac{r}{r_{\text{int}}}\right) + \frac{1}{r_{\text{int}} \alpha_{\text{int}}} \right] +$$
$$+ \sum_{j=1}^{\infty} A_{1j} \cos \xi \left[I_0(\rho) + B_1 K_0(\rho) \right] \tag{2.104}$$

It can be concluded from Eq. (2.104) that the temperature field within the wall of a finned tube in the axial direction consists of constant and a periodic parts. The constant part, in its turn, also consists of two components, namely the temperature gradient which would occur in a bare wall if $\bar{\alpha}$ is taken as the heat transfer coefficient, plus a term designating the reduction of temperature level within the tube wall. Both parts depend directly on the difference $\alpha^* - a_0$. When this difference vanishes, the temperature-field equation (2.104) becomes an equation for a bare tube with constant heat transfer conditions. In this respect the heat transfer coefficient α_0 of the tube to which the fins are attached is regarded as very important.

In a very thin tube wall ($r_{\text{int}} \to r_0$) the radial temperature gradient vanishes and only the axial gradient ϑ_z remains. Also, in this limiting case of an infinitesimally thin tube wall there is no conduction in axial direction. Then, the temperature ϑ_1 of the wall at the fin base and between the fins ϑ_1' can be expressed in the form:

$$\vartheta_1 = \frac{\vartheta_{f1}}{\alpha_{\text{int}} \left(\frac{1}{\alpha^*} + \frac{1}{\alpha_{\text{int}}} \right)}, \qquad \vartheta_1' = \frac{\vartheta_{f1}}{\alpha_{\text{int}} \left(\frac{1}{\alpha_0} + \frac{1}{\alpha_{\text{int}}} \right)} \tag{2.105}$$

CONDUCTION OF HEAT IN A FINNED TUBE

The mean tube wall temperature in this case is expressed as:

$$\vartheta_m = \frac{\delta}{s}\vartheta_1 + \left(1 - \frac{\delta}{s}\right)\vartheta'_1 \tag{2.106}$$

or, according to Eq. (2.95)

$$\vartheta_m = \frac{\vartheta_{f1}}{\alpha_{int}\left(\frac{1}{\bar{\alpha}} + \frac{1}{a_{int}}\right)}\left[1 - \frac{2\left(\frac{\alpha^* - \alpha_0}{\bar{\alpha}}\right)^2}{(\bar{\alpha} + \alpha_{int})\left(\frac{1}{\bar{\alpha}} + \frac{1}{\alpha_{int}}\right)}\sum_{j=1}^{\infty}\frac{\sin^2\bar{\delta}}{(j\pi)^2}\right] \tag{2.107}$$

If the tube is thick-walled and has closely spaced fins, then the periodic temperature gradient gradually equalizes within the wall and, at a certain depth, the temperature no longer depends on the axial position. The definition of a thick tube wall is

$$\rho_0 - \rho_{int} = 2\pi j\frac{r_0 - r_{int}}{s} > 3, \qquad \frac{r_0 - r_{int}}{s} > 0.5 \tag{2.108}$$

According to Krischer and Kast [39], the actually observed difference in heat flux through the fin region and through the base tube region penetrates to a position $r \simeq r_0 - 0.5s$. For such a thick tube wall the periodic term in Eq. (2.104) no longer depends on α_{int}. This means that the temperature on the inner tube wall ($r = r_{int}$ is constant and is independent of the axial position.

The temperature gradient in the thick tube wall for $r_{int}/s > 1.2$ and $(r_0 - r_{int})/s > 0.5$ can be obtained from the simplified equation:

$$\vartheta(r, z) = \frac{\vartheta_{f1}}{\frac{1}{r_0\bar{\alpha}} + \frac{1}{\lambda}\ln\left(\frac{r_0}{r_{int}}\right) + \frac{1}{r_{int}\alpha_{int}}} \times$$

$$\times\left\{\left[\frac{1}{\lambda}\ln + \frac{1}{r_{int}\alpha_{int}}\right]\cdot\left[1 - \frac{\left(\frac{\alpha^* - \alpha_0}{\bar{\alpha}}\right)^2\frac{s}{r_0}\sum_{j=1}^{\infty}\frac{\sin^2\bar{\delta}}{(j\pi)^3}}{\lambda\left[\frac{1}{r_0\bar{\alpha}} + \frac{1}{\lambda}\ln\left(\frac{r_0}{r_{int}}\right) + \frac{1}{r_{int}\alpha_{int}}\right]}\right]+\right.$$

$$\left. + \frac{\alpha^* - \alpha_0}{\bar{\alpha}\lambda}\frac{s}{r_0}\sum_{j=1}^{\infty}\frac{\sin\bar{\delta}}{(j\pi)^2}\frac{I_0(\rho)}{I_0(\rho_0)}\cos\xi\right\} \tag{2.109}$$

To obtain a clear idea about the variation in temperature gradient within the tube wall, Krischer and Kast [39] used the aforementioned equations for calculat-

46 HEAT TRANSFER OF FINNED TUBE BUNDLES IN CROSSFLOW

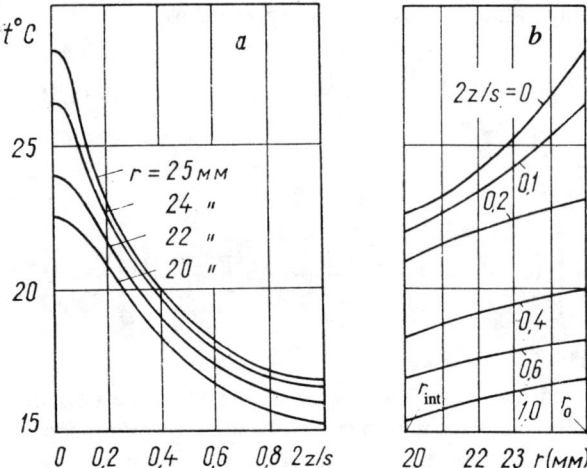

Figure 2.9 Temperature gradient in the wall of a thin-walled tube. *a*) In axial direction, as a function of nondimensional coordinate $2z/s$ at different r; *b*) in radial direction, as a function of r at different $2z/s$.

ing the gradient for a thin-walled finned tube with the following dimensions: $r_{int} = 20$ mm, $r_0 = 25$ mm, $\delta = 5$ mm, $s = 50$ mm, $\lambda = 20$ W/m × K. Figure 2.9 shows the results of these calculations in axial and radial directions of the tube. It is seen that the temperature gradient varies significantly more in the z direction.

Calculations for a thick-walled finned tube were performed for the dimensions: $r_{int} = 45$ mm, $r_0 = 85$ mm, $\lambda = 3$ mm, and $s = 14$ mm. It is seen from the results shown in Fig. 2.10 that in this case the radial temperature gradient is the controlling factor.

To illustrate the above, the results of calculations are plotted in Fig. 2.11, in the form of temperature isotherms within the wall for the respective cases. Consider now the temperature changes along the tube wall we see that, for the internal tube surface, it is important to establish when this longitudinal gradient becomes zero, i.e., when $\Delta\vartheta_{int} = 0$. It was established for the case of $r_{int} > 1.2s$ that this is attained at wall depth $r_0 - r_{int} > 0.5s$.

The longitudinal temperature gradient at the outer surface of the finned tube between $z = 0$ and $z = s/2$, i.e., $\Delta\vartheta_1 = \vartheta(r_0, 0) - \vartheta(r_0, s/2)$ and for $r - r_{int} > 0.5s$ and $r_{int} > 1.2s$ can be calculated from the equation

$$\Delta\vartheta_1 = \frac{2\vartheta_{f1}}{\frac{1}{r_0\bar{\alpha}} + \frac{1}{\lambda}\ln\left(\frac{r_0}{r_{int}}\right) + \frac{1}{r_{int}\alpha_{int}}} \cdot \frac{\alpha^* - \alpha_0}{\bar{\alpha}\lambda} \sum_{j=1,3,5\ldots}^{\infty} \frac{s}{r_0} \frac{\sin\bar{\delta}}{(j\pi)^2} \quad (2.110)$$

Figure 2.12 shows, in nondimensional variables, a plot of $\Delta\vartheta_1$ vs fin thickness

CONDUCTION OF HEAT IN A FINNED TUBE 47

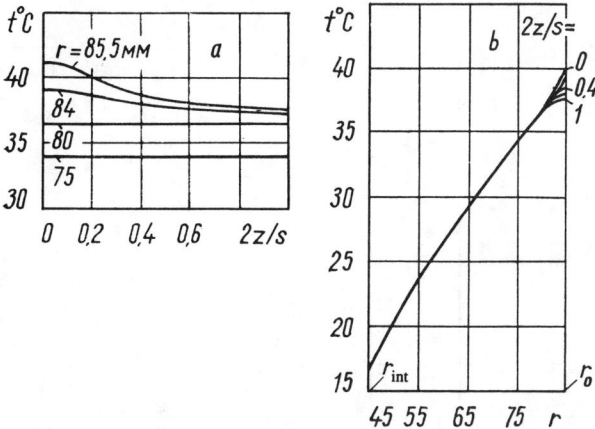

Figure 2.10 Temperature gradient in the wall of a thick walled tube. *b*) In axial direction, as a function of $2z/s$ at different r; *b*) in radial direction as a function of r at different $2z/s$.

δ/s as calculated from Eq. (2.110). The parameter here is α_0/α^*. It is seen that $\Delta\vartheta_1$ takes rapidly with increasing α_0/α^*, the maximum of the curves being shifted here in the direction of smaller δ/s. This occurs because since high heat fluxes passing through small fin base cross sectional areas will involve the appearance of a higher temperature difference $\Delta\vartheta_1$, than for low heat fluxes passing through large base cross sectional areas.

Let us now determine the error in calculating the heat transfer of a finned tube using the simplification described previously, i.e., assuming that the finned tube can be treated as a plain tube with an axially constant heat transfer condition but with a heat transfer coefficient $\bar{\alpha}$ calculated using an area corresponding to a cylinder of radius r_0 and enhanced due to the presence of the fins. The overall heat transfer coefficient, k_1, referred to unit tube length, can be written as:

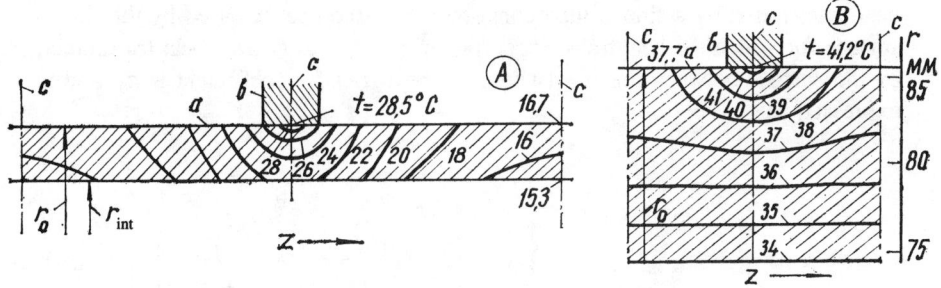

Figure 2.11 Temperature field in the wall of a finned tube. *a*) Wall of finned tube; *b*) fin; *c*) axial lines. *A*) thin-walled tube; *B*); thick-walled tube.

48 HEAT TRANSFER OF FINNED TUBE BUNDLES IN CROSSFLOW

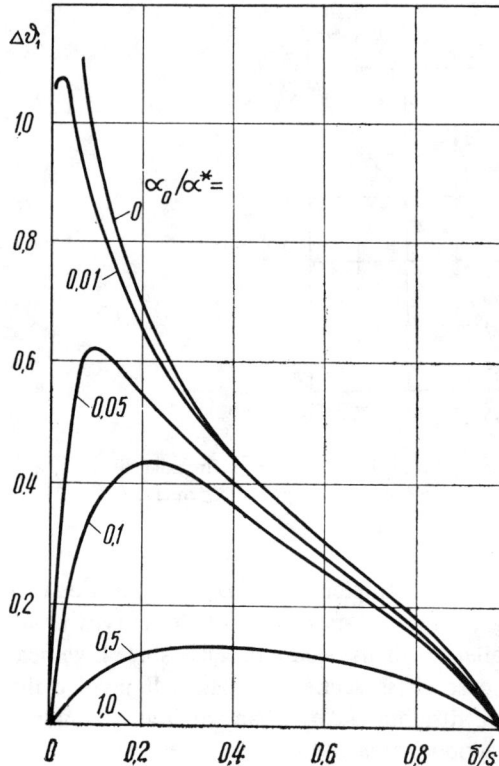

Figure 2.12 Nondimensional temperature difference along the outer surface of a finned tube

$$\Delta\vartheta = \Delta\vartheta_1 \lambda r_0 (\vartheta_{f1} - \vartheta_{f2}) k_1 s$$

as a function of δ/s at different α_0/α^*.

$$k_1 = \frac{1}{\dfrac{1}{\bar{\alpha} r_0} + \dfrac{1}{\lambda} \ln\left(\dfrac{r_0}{r_{int}}\right) + \dfrac{1}{\alpha_{int} r_{int}}} \qquad (2.111)$$

The amount of heat transmitted by such a tube is obtained from Eq. (2.75). The heat transmitted by a finned tube consists of Q_{fin}, the heat released by the finned surface for which the heat transfer coefficient is α^*, plus Q_0, the heat transmitted by the surface between the fins (where the heat transfer coefficient is α_0) and is expressed as:

$$Q = Q_{fin} + Q_0 = 2\pi r_0 l \left\{ \frac{1}{s} \int_{-1/2\delta}^{1/2\delta} \alpha^* [\vartheta_{f1} - \vartheta] (r_0, z) \, dz + \right.$$

$$\left. + \frac{1}{s} \int_{1/2\delta}^{s-1/2\delta} \alpha_0 [\vartheta_{f1} - \vartheta (r_0, z)] \, dz \right\} = 2\pi l K' \vartheta_{f1} \qquad (2.112)$$

where it is assumed that $\vartheta_{f2}=0$, where k_0 is the heat transfer coefficient of the tube carrying the fins.

According to Eqs. (2.104) and (2.111), we find:

$$k_0 = k_1 \left[1 - 2K_0' \left(\frac{\alpha^* - \alpha_0}{\bar{\alpha}} \right)^2 \frac{1}{r_0 \bar{\alpha}} \sum_{j=1}^{\infty} \frac{\sin^2 \bar{\delta}}{(j\pi)^2} \frac{1}{M_j} \right] \quad (2.113)$$

The heat transfer coefficient k_0 will be smaller than its analogous coefficient k_1 for a circular tube (with mean heat transfer coefficient $\bar{\alpha}$). The difference $k_1 - k_0$ is a function of a number of factors, primarily of k_1, $\alpha^* - \alpha_0$ and δ/s.

Analysis of the above shows [39] that $(k_1 - k_0)$ increases with increasing $(\alpha^* - \alpha)$ and with increasing δ but decreases with increasing s. [This applies also to the temperature difference $\Delta\vartheta_1$ on the outer surface of the tube (see Fig. 2.12).]

In the most frequently encountered tubes (with $\alpha_0/\alpha^* < 0.2$) $\Delta k = k_1 - k_0$ does not exceed 1%. Hence, the majority of real cases the reduction Δk can be neglected, and the previously stated expressions can be used.

CHAPTER
THREE

FLOW OVER AND HEAT TRANSFER OF FINNED TUBE BUNDLES

In Chapter 2 we presented solutions which take account of the effects of fin shape, thickness, height, and thermal conductivity of temperature and heat flux over the fin surface. These solutions were obtained from analytic solution of the problem of heat conduction in a finned tube; they make no allowance whatsoever for the conditions of flow over the finned surface and the resultant temperature distribution over the fins. For this reason the practical application of these solutions for calculating the heat transfer from a finned tube, and the more so from bundles of such tubes, is limited.

Below we consider the principal problems of convective heat transfer from bundles of finned tubes.

3.1. FACTORS CONTROLLING THE TRANSFER OF HEAT FROM A FINNED TUBE BUNDLE

The conditions of flow and the distributions of temperature fields within bundles of finned tubes (which comprise a set of complex-shaped bodies in external flow) are complex and highly variagated, since they depend on a large number of thermal, hydraulic and other factors. Thus, heat transfer from a bundle of finned tubes depends on a large number of geometric and other parameters.

If a smooth and plane heated surface is placed in a uniform coolant flow, the effect of the body's shape on heat transfer and hydraulic drag is represented by a small number of parameters, controlled by the body's geometry. These parameters, in the case of some surfaces—for example a smooth plate or the inner surface of a tube—placed in an axial flow, control the variation of velocity and temperature fields along the body, and also the variation in the thickness of the boundary layer. In the case of a plane finned surface in axial flow, a larger number of parameters is needed in order to describe the flow pattern and the heat transfer. In addition to the length of flowpath along the body, it is at least necessary to have information on the height and pitch of the fins. Flow over such a surface along the fin can be approximately taken as being identical to flow over a plane plate, and hence the variation in the coefficient of heat transfer along the fin can be estimated by simple relationships, which are different for laminar and turbulent flow regions.

The velocity and temperature distributions about a body in transverse flow are more complicated. They depend on flow conditions upstream of the body, on the flow over the perimeter of the body to the point of separation, and on vortex generation in the body's wake. Hence heat transfer from a body in cross flow is usually subdivided into three circumferential zones, the conditions within each expressed by different equations. The complexity of the situation is additionally seen from the fact that even for geometrically simple, single bodies, such as, for example, a circular tube, described by dimensions d and l, the temperature and velocity distributions over the perimeter can be described only in the zone upstream of the point of boundary-layer separation [33].

Transverse flow over a bundle of bodies is an even more complex process. The inherent complexity of flow over a bundle of finned tubes can be gleaned, among others, from visual observations by Neal and Hitchcock [57], performed with a model of nine-row staggered bundle at $Re = 1.25 \cdot 10^5$.

According to these observations the conditions of flow over the front and deeper-lying rows differ significantly. The flow over the first two rows is even, but then as the number of rows along the flow direction increases, ever increasing levels of turbulence are found in the spaces between the tubes and between the fins. The overall flow pattern over a finned tube, characteristic of both the front and also the deeper lying rows, are shown in Fig. 3.1.

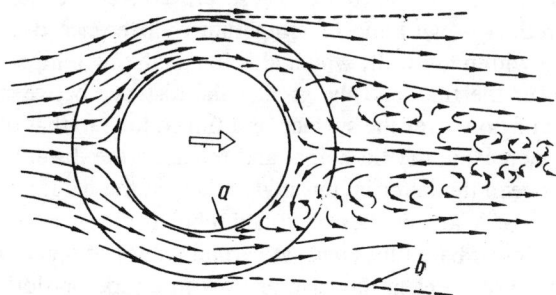

Figure 3.1 Pattern of flow over a finned tube. *a)* Point of flow separation; *b)* approximate boundaries of elevated turbulence.

As can be seen, flow separation occurs on the downstream half of the tube perimeter. Separation is followed by a wedge-shaped vortex zone. There is a strong re-entrant jet along the axis of the zone, acting on the finned tube in a manner similar to a free stream. The re-entrant jet flows over the finned-tube surface up to the point of mainstream separation. The elevated vorticity and turbulence of the flow in the center of the wedge rapidly diffuse into the mainstream. The turbulence of the wedge-shaped stream increases rapidly with distance into the bundle and, approximately at the fourth row, it starts to strongly affect the entire separating flow.

Comparison of the nature of flow over the second and sixths rows shows that they differ somewhat: the flow over the sixth row is highly turbulized, the point of separation on the tube of the sixth row is shifted somewhat down-steam (as compared with a tube in the second row), the wedge-shaped flow is narrower and the re-entrant jet has a more complex configuration.

An indicator commonly used for heat exchangers is the ratio between the heat transfer and drag:

$$\lambda = \frac{8 \text{St}}{\xi} \tag{3.1}$$

According to theory of similitude, for undisturbed flow $\lambda = 1$ [42]. For tubes and flat shots with constant cross section for the coolant flow, the value of λ approaches this theoretical value with increasing Re. The value of λ for circular and finned tubes in crossflow deviates significantly from the theoretical value. This deviation is greater in bundles of bare than of finned tubes. This is due to the fact that the pressure drop expended for overcoming the changes in cross section and the vortex generation in these bundles, exceed manyfold the friction losses.

In the majority of cases the heat transfer coefficient is obtained as a mean value, calculated from the total amount of transmitted heat and the temperature difference between the mean flow and wall temperatures. The heat transfer coefficient calculated in this manner is an average affected by many different phenomena, occurring on the surface of the body. These are convective heat transfer and flow, (which produce a variation of temperature difference over the body) and heat conduction within the wall, which aids in equalizing temperature difference, etc. The lower the fins, and the greater the distance between them, the closer is the pattern of flow over the surface of a finned tube to that of flow past a bare tube. Upon increasing the fin height and reducing the distance between the fins, the flow between the latter increasingly takes on the nature of flow in a slot. Hence the value of λ for bundles of finned tubes is closer to the theoretical value than for bundles of bare tubes and, at certain values of h and u it can be rather close to values prevalent in the case of flow in a pipe or slot.

Data presented by Weiner, et. al.[55] and Neal and Hitchcock [57] on the distribution of the heat transfer coefficient on the surface of the tube and fin, point to the existence of a large difference between them. This difference depends, in addition to everything else, on the conditions of flow approaching the tube. In addition, heat transfer from a tube in a bundle depends on the distribution of tubes, and also on the row, in the direction of flow, in which the given tube is situated.

This means that heat transfer from a finned tube in a bundle can, in nondimensional form, be described by the following functional relationship:

$$\text{Nu} = f\left(\text{Re; Pr; } \frac{T_f}{T_w}; \frac{\lambda_f}{\lambda_w}; o_1; o_2; o_3; \frac{s_1}{d}; \frac{s_2}{d}; z\right) \quad (3.2)$$

where o_{1-3} are parameters estimating the finning geometry. The flow mode is described by the Reynolds number, and the physical properties of the coolant are described by the Prandtl number. The effect of these properties on heat transfer is closely related to their variation in the boundary layer. The ratio of flow to wall temperatures T_f/T_w (the so-called temperature factor) is an indicator estimating the effect on heat transfer of the variability of the physical properties of the coolant (μ, λ, c_{fin} and ρ) in the boundary layer. The effect of this factor, at least over the temperature range actually encountered in practice, is small in the case of a gaseous coolants.

It is appropriate to mention the effect on heat transfer of flow conditions at the bundle inlet, including also the turbulence of the free steam. According to available data, heat transfer in a tube bundle (particularly in the first rows) depends highly on the mainstream turbulence, which in many cases depends directly on the design of the inlet part of the experimental arrangement.

3.2. COEFFICIENT OF HEAT TRANSFER FROM A FINNED TUBE

The introduction of the fin effectiveness in calculating heat transfer from finned tubes makes it possible to forego the use of rather complicated analytic expressions and to perform calculations in a form simpler and more suitable for practical use.

Thus, if we assume that the temperature of the liquid surrounding the finned tube is equal to zero, then the amount of heat transmitted by the fin surface, according to Eq. (2.64), is:

$$Q_{\text{fin}} = \alpha t_l E F_{\text{fin}} \quad (3.3)$$

Accordingly, the heat given up by the smooth part of the finned tube

$$Q_t = \alpha_0 t_l F_t \quad (3.4)$$

Then the total amount of heat can be expressed as:

$$Q = Q_{\text{fin}} + Q_t = \alpha_1 E F_{\text{fin}} + \alpha_0 t_1 F_t \qquad (3.5)$$

If the heat transfer coefficient is represented by its mean value α_{red}, which makes allowance for heat transfer from the fin surface and from the surface of the part of the tube without fins, and also the effectiveness of the fin performance, then the total amount of heat released by the finned surface is:

$$Q = \alpha_{\text{red}} t_1 F \qquad (3.6)$$

where

$$\alpha_{\text{red}} = \alpha E \frac{F_{\text{fin}}}{F} + \alpha_0 \frac{F_t}{F} \qquad (3.7)$$

$$\frac{F_t}{F} = 1 - \frac{F_{\text{fin}}}{F} \qquad (3.8)$$

Assuming that $\alpha_0 \simeq \alpha$, we obtain:

$$\alpha_{\text{red}} = \alpha \left[\frac{F_{\text{fin}}}{F} E + \left(1 - \frac{F_{\text{fin}}}{F}\right) \right] = \alpha \left[1 - (1 - E) \frac{F_{\text{fin}}}{F} \right] \qquad (3.9)$$

Coefficient α_{red} is extensively used in calculating heat transfer from finned surfaces. In Soviet literature this coefficient is termed "reduced," in German literature the term is "scheinbare" (apparent) [21]. In American literature this coefficient frequently does not have a specific name; in defining the coefficient of heat transfer from a finned surface the term in brackets in Eq. (3.9) is termed the "efficiency" [23]. Sometimes it is termed the "effective surface coefficient" [74].

Coefficient α, pertaining to the fin surface, is termed in the Soviet literature the "convective" coefficient. In Western literature one frequently encounters terms such as "wahr" (German) "true."

The "reduced heat transfer coefficient" α_{red} implicitly includes the thermal resistance to conduction due to the shape, cross section and material of the fin, and the thermal resistance to convective heat transfer to the coolant passing over the finned surface, including also the heat transfer effectiveness of the fin. α_{red} is defined as:

$$\alpha_{\text{red}} = \frac{Q}{F \Delta t} \qquad (3.10)$$

where the heat transfer surface is understood to be either the total surface F of the finned tube, or the surface F_1 a plain tube whose diameter is equal to the fin root diameter. In the first case the absolute values of α_{red} are lower than α_1, in the second they are higher.

According to Grass and Coenen [40], α_{red} can be expressed as

$$\alpha_{red} = \varepsilon_F \varepsilon_L \varepsilon_u \varepsilon_\alpha \alpha_1 = \epsilon_{fin} \alpha_1 \qquad (3.11)$$

or

$$\alpha_{red} = \varepsilon_F \varepsilon_L \varepsilon_u \alpha$$

where ϵ_F is the surface extension factor (corresponding to the factor φ), ϵ_L is a factor representing the effect of temperature variation over the fin height (the fin effectiveness), ϵ_u is a coefficient accounting for the degree of contact between the fin and the tube (in cases when these are not integral), $\epsilon_\alpha = \alpha/\alpha_1$ is a coefficient accounting for the change in the convective heat transfer coefficient of a finned tube relative to one without fins, and $\epsilon_{fin} = \alpha_{red}/\alpha_1$ is a ratio accounting for the overall change in the heat transfer coefficient as compared to that for a plain tube.

Increasing the surface of a tube by attaching fins to it (by a factor ϵ_F) does not increase the amount of heat transferred per unit area, i.e., $\epsilon_{fin} < \epsilon_F$. The ratio $\epsilon_{fin}/\epsilon_f$ characterizes the overall increase in heat transfer, which could be obtained for a finned tube with given dimensions in the case of infinite thermal conductivity. From this relationship, and for $\epsilon_u = 1$, we can define, on the basis of Eq. (3.11), the value of ϵ_α from the expression

$$\varepsilon_\alpha = \frac{\epsilon_{fin}/\epsilon_L}{\varepsilon_F} \qquad (3.12)$$

ϵ_α characterizes the change in the convective heat transfer coefficient compared with that for the plain tube.

As an illustration, Fig. 3.2 depicts the variation of ϵ_α for a straight circumferential fin as a function of fin height, fin spacing and Re; this data was obtained experimentally for a single finned tube with fin base diameter $d = 30$ mm [40]. It is seen from the graphs that the convective coefficient of heat transfer from a finned tube increases (as expected) with the Re but decreases with increasing fin height and reducing fin pitch.

The effect on effectiveness coefficient of having trapezoidal rather than rectangular fin cross section shape (as frequently occurs in practice) can be accounted for by a correction factor ξ, which can be determined from the graph in Fig. 3.3 as a function of parameters βh and δ_2/δ_1 [15].

It must be remembered that in determining an average value of α from, say, the measured α_{red} values, proper attention must be paid to the fin effectiveness, E; serious errors occur when this is not done. In determining the value of α using Eq. (3.9), errors may occur due to errors in the experimental determina-

56 HEAT TRANSFER OF FINNED TUBE BUNDLES IN CROSSFLOW

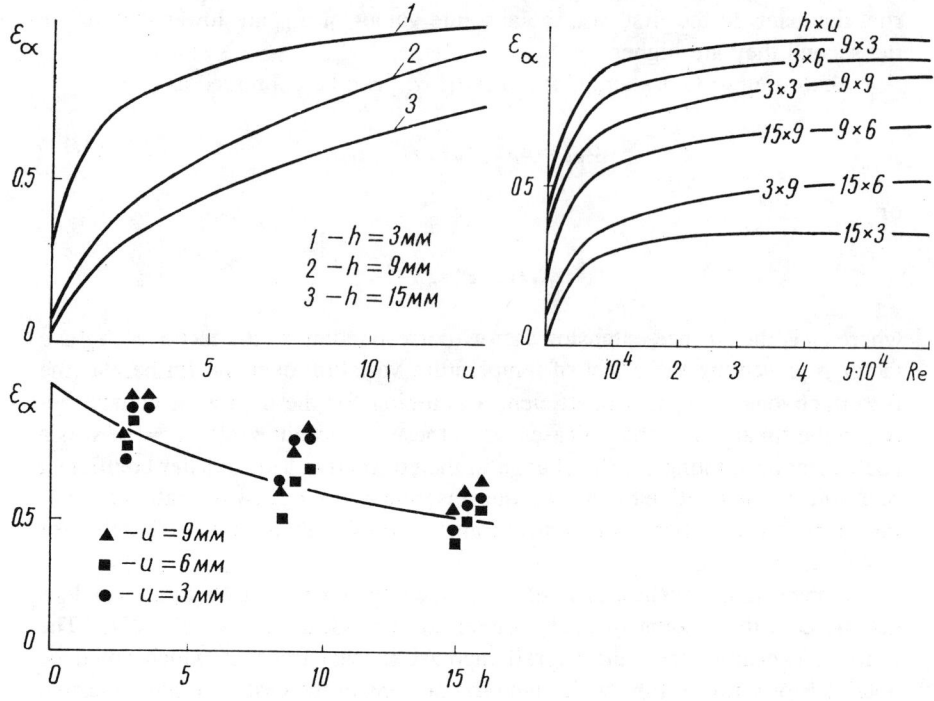

Figure 3.2 Coefficient ϵ_α, vs Re, spacing between the fins and the height of a straight circumferential fins.

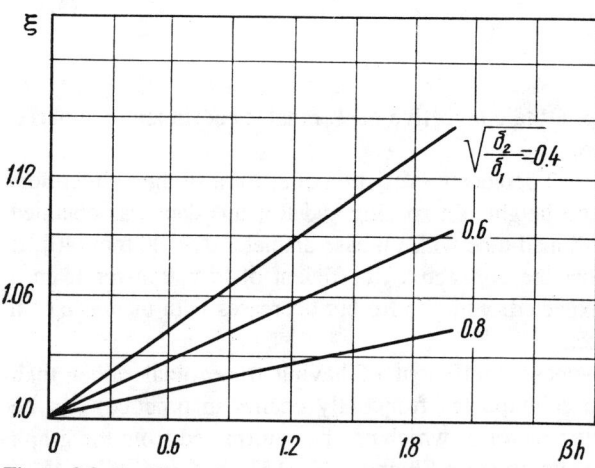

Figure 3.3 Coefficient ξ, correcting for the variation of thickness of trapezoidal fins.

tion of α_{red}, and also inaccuracies in determining E. Furthermore, the interaction between local variations in coefficient on the fin surface and conduction within the fin may make the average α value different. The scale of the effect of the fin effectiveness on α for a given α_{red} becomes clearer if Eq. (3.9) is written in the form suggested by Schmidt [37]:

$$\alpha_{red} = \alpha E \left(1 - \frac{u}{s} \frac{F_1}{F} \frac{1-E}{E}\right) \qquad (3.13)$$

In the majority of cases, and particularly at high E, the second term in parentheses is small as compared with unity. This means that a change in α_{red} will result in an almost equivalent change in αE.

It will be seen by carefully examining Eq. (2.72) that, since $\alpha E = X^2 E = X\tanh X$, the relative variation in α as a function of α_{red} can be determined from Fig. 3.4. It can be easily found from the graph that, at high values of E (for example, for $E = 0.85$ corresponding to $X = 0.74$ and $X\tanh X = 0.465$) a 10% deviation in α_{red} changes α by 12%. At low E, and respectively high X, we find that $X\tanh X \to 1$, then

$$\frac{1-E}{E} \to (X-1) \qquad \qquad 68A$$

Obviously, in this case the second term in parenthesis in (3.13) can no longer be neglected, and errors in determining α_{red} will result in larger errors in α. Equa-

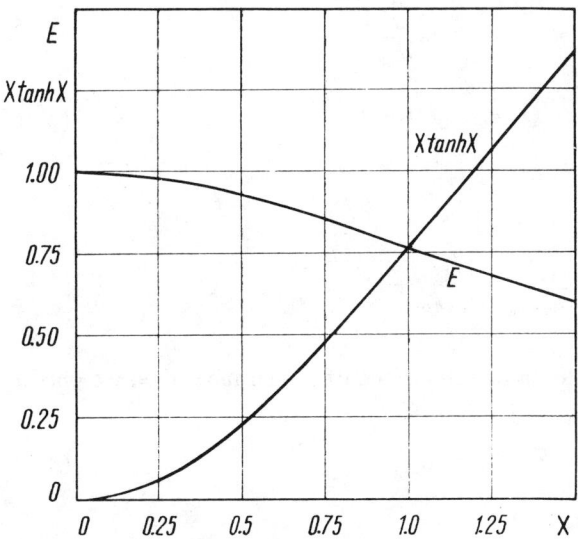

Figure 3.4 Plot of E and $X\tanh X$ vs X.

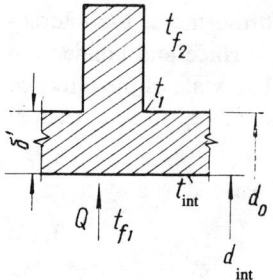

Figure 3.5 Schematic of finned tube.

tion (3.13), relating the heat transfer coefficient to a number of geometric parameters of the fin is of interest in practical determination of the mean heat transfer coefficient of a fin, in spite of the fact that it lacks sufficient analytic basis.

It must be additionally noted that the fin effectiveness also affects the value of m in Eq. (1.1). Thus, for example, E decreases with increasing Re (in conjunction with more rapid cooling or heating of the fin), and failure to correct for this will lead to an underestimate of the value of m. Here the effect of E on the value of m increases with decreasing E.

In calculating the rate of heat transfer through a finned tube between two fluids f_1 and f_2 (see Fig. 3.5), we can write the following equations:

$$Q = \alpha_{int} F_{int}(t_{f1} - t_{inf}) \quad (3.14)$$

$$Q = \frac{2\pi l \lambda (t_{inf} - t_1)}{\ln \frac{d_0}{d_{inf}}} \quad (3.15)$$

$$Q = \alpha_{red} F(t_1 - t_z) \quad (3.16)$$

These equations yield

$$Q = \frac{t_{f1} - t_{f2}}{\dfrac{1}{\alpha_{int} F_{int}} + \dfrac{1}{2\pi l \lambda} \ln \dfrac{d_0}{d_{int}} + \dfrac{1}{\alpha_{red} F}} \quad (3.17)$$

If the heat flux is referred to unit total surface of finned tube, then we obtain

$$\frac{Q}{F} = q = k\,(t_{f1} - t_{f2}) \quad (3.18)$$

where

$$k = \cfrac{1}{\cfrac{1}{\alpha_{int} F_{int}} + \cfrac{F}{2\pi l \lambda} \ln \cfrac{d_0}{d_{int}} + \cfrac{1}{\alpha_{red} F}}$$

If, however, the heat flux is referred to unit surface F_1 of the tube carrying the fins, then

$$\frac{Q}{F_1} = q_1 = k_1 (t_{f1} - t_{f2}) \qquad (3.19)$$

where

$$k_1 = \cfrac{1}{\cfrac{1}{\alpha_{int}} \cfrac{F_1}{F_{int}} + \cfrac{F_1}{2\pi l \lambda} \ln \cfrac{d_0}{d_{int}} + \cfrac{1}{\alpha_{red}} \cfrac{F_1}{F}}$$

In the majority of practical cases the tube wall thickness is small compared with the tube diameter. In this case calculations are performed with simplified formulas. Assuming that $d_0/d_{int} \to 1$, and expanding $\ln(d_0/d_{int})$ in a series, and then retaining only its first term, we obtain:

$$\ln \frac{d_0}{d_{int}} = \frac{2\delta'}{d_{int}} \qquad (3.20)$$

Assuming then that $F_{int} \simeq F_{int}$, we obtain simplified formulas for heat transfer coefficients for a finned tube for both cases:

$$k_1 = \cfrac{1}{\cfrac{1}{\alpha_{int}} \cfrac{F}{F_1} + \cfrac{\delta'}{\lambda} \cfrac{F}{F_1} + \cfrac{1}{\alpha_{red}}} \qquad (3.21)$$

$$k_1 = \cfrac{1}{\cfrac{1}{\alpha_{int}} + \cfrac{\delta'}{\lambda} + \cfrac{1}{\alpha_{red}} \cfrac{F_1}{F}} \qquad (3.22)$$

3.3. SELECTION OF THE PRINCIPAL PARAMETERS

It is seen from Eq. (3.2) that reduction of experimental data on heat transfer from a finned tube bundle involves the use of a large number of parameters and reference dimensions. Different investigators employ different sets of these variables, which frequently complicates the comparison of results.

1. Linear Reference Dimension

Nondimensional similitude criteria must be based on a reference linear dimension. Up to now there has been no consensus on the best reference dimension to be used for finned-tube bundles. With reference to convenience of calculation many investigators take the diameter d of the tube carrying the fins as the reference dimension. Others use for this purpose the equivalent diameter

$$d_{eq} = \frac{4f}{U} \tag{3.23}$$

where f is the flow cross section of the bundle (the shaded area in figure 3.6), whereas U is the wetted perimeter (indicated by the bold line in Fig. 3.6). The latter is more convenient to use in heat exchangers with cross sectional area f constant in the direction of coolant flow and the case when the velocity distribution over the cross section is primarily controlled by friction. According to Schmidt [37], d_{eq} for finned tube bundles should be expressed as:

$$d_{eq} = \left[\frac{4}{\pi} \frac{s_1}{d} \frac{s_2}{d} - 1\right] \frac{F_1}{F} d \tag{3.24}$$

Sometimes the fin pitch s is used as the reference dimension.

Krischer [39] uses the streamline length l'. This is the mean path, traversed by a stream particle flowing over the body. For a circular tube it is expressed in the form

$$l' = \frac{\pi}{2} d \tag{3.25}$$

For a tube with circumferential fins

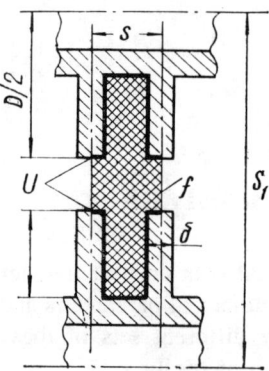

Figure 3.6 For determination of the equivalent diameter.

$$l' = \frac{\pi}{2} d \sqrt{1 + \left(\frac{h}{d}\right)^2} \tag{3.26}$$

and for a tube with spiral fins

$$l' = \frac{\pi}{2} d \left(1 + \frac{h}{d}\right) \tag{3.27}$$

It is seen that the last two expressions for l' contain parameter h/d, which, as shall be seen below, is an important finning parameter.

Schmidt [37] considered, as a part of his analysis of the linear reference dimension, the surface extension factor φ. The latter can be expressed as

$$\varphi = \frac{F_1}{F} = \frac{F/L}{\pi d} \tag{3.28}$$

where L is the length of the tube. The ratio F/F_1 is a factor giving the increase in the surface area of a finned tube, as compared to a bare one. The denominator in Eq. (3.28) is twice the streamline length of a nonfinned tube.

Schmidt suggested that the reference dimension be identified with the so-called equisurface diameter

$$d_F = d \frac{F}{F_1} \tag{3.29}$$

The advantage of using this diameter is the fact that both the nonfinned and finned tube with the same outer surface will have a diameter defined by Eq. (3.29). The use of d_F in the Reynolds number, as compared, for example, with d_{eq}, results in the following effect: when the value of Re is based on d_{eq}, Re decreases with increasing φ; conversely, when the value of Re is based on d_F, Re increases with φ.

2. Parameters Estimating the Finning Geometry

A large number of opinions have been put forward concerning the use of certain nondimensional geometric finning parameters in experimental data reduction.

For example, Antuf'yev [8] uses the ratios h/d, u/d and δ/d. Note that ratio δ/d has little effect on α_{red}. Experimental studies show that a five-fold reduction in fin thickness results in a mere 13% reduction in the heat transfer from a finned tube in a staggered bundle. True, the experiments were performed with fins with $h/d = 0.33$. Since the thermal resistance increases with the fin height, the effect of fin thickness on α_{red} may be greater with higher fins. (The heat transfer coefficient used in these studies referred to the surface of the tube carrying the fins.). Analysis of the effect of u/d shows that α_{red}, referred to the surface of a bare tube, decreases with increasing u/d; however, the value of m in Eq. (1.1) remains constant. Antuf'yev estimated that heat transfer coefficient varied as $(u/d)^{-0.8}$ for staggered bundles, and as $(u/d)^{-0.5}$ for in-line bundles. In an in-line bundle, because of the screening effect of the tubes of the preceding row, the velocity of the gas at the fin base is lower than in a staggered bundle, for which reason the effect of u on heat transfer is smaller Heat transfer coefficient increases with increasing h/d, whereas the exponent m (Eq. 1.1) decreases perceptibility due to the rising thermal resistance. It was found that for staggered bundles $m = 0.69\text{--}6.37\ h/d$ whereas for in-line bundles $m = 0.78\text{--}0.35\ h/d$. For the range $0.167 < h/d < 0.25$, the heat transfer coefficient was found to be proportional to $(h/d)^{1.16}$ for staggered bundles, whereas for in-line bundles it is proportional to $(h/d)^{1.36}$. For higher fins, the exponents of this ratio are 1.98 and 1.75, respectively.

Karasina [13] employs ratios d/s and h/s as geometric functions of the finning. Their effect on heat transfer (determined on the basis of the convective heat transfer coefficient α) in in-line and staggered bundles was found to be proportional to $(d/s)^{-0.54}$ and $(h/s)^{-0.14}$, respectively.

Krischer and Kast [39], in investigating heat transfer from a finned tube, pay particular attention to the reduction in stream velocity between the fins. The latter is controlled by the geometric parameters of the fin, which are h/u, h/d and u/s. The higher h/u, i.e., the higher the fins and the closer they are to one another, the greater the friction loss in the flow between the fins. Finally, ratio u/s, $= -\delta/s$ reflects the effect of fin thickness on heat transfer. At low fin thicknesses (δ) the flow between the fins undergoes only moderate changes (it narrows down somewhat at the inlet). In the limiting case of $\delta \rightarrow 0$ the mean flow velocity w_m before entering the space between the fins is equal to the mean velocity in this latter space w_{fin}, i.e., $\omega_m s = \omega_{fin}(s-\delta)$, or $\omega_{fin}/\omega_m = (1-\delta/s)_{-1}$.

Figure 3.7 shows the effect of the aforementioned parameters on the exper-

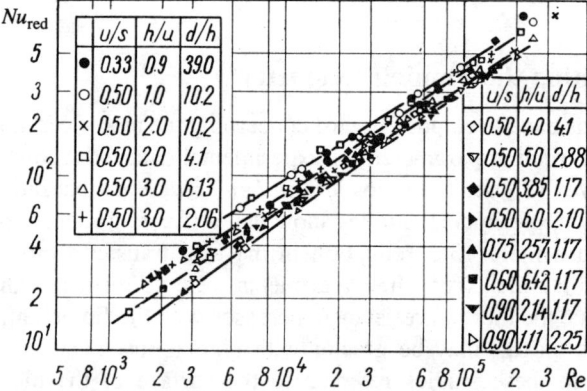

Figure 3.7 Effect of finning geometry on heat transfer from a finned tube.

imentally determined heat transfer coefficient, plotted as a function of Re [39]. The effect of the parameter h/u is most significant. The higher this ratio, i.e., the narrower and higher the stream cross section between the fins, the lower the experimental points. (In [39] the reference dimension in Nu_{red} and Re is the streamline length l'.). It is seen that the effect of these parameters is more significant at low Re.

Brauer [17], in analyzing the effect of geometric parameters of finning on heat transfer, assumes that the surface extension factor φ be used as the independent variable, rather than u and h. He points out, in comparing the heat transfer from a bundle of finned and that of a bundle of bare tubes, expressed as $\text{Nu}_{\text{red}}/\text{Nu}_0 = f(\varphi)$ at different values of u, that for a bundle with closely spaced tubes (for example, $u = 2$ mm), an initial moderate increase in φ does not increase the amount of heat transferred. This points to the existence of a high hydraulic resistance. The thick boundary layer in this case is accomplished by a high hydraulic drag coefficient. Brauer concludes from this that moderate increase in φ should be obtained by increasing the fin height, rather than by increasing the fin density, since the latter decreases the distance between the fins with attendant rise in hydraulic drag.

Schmidt [37] notes that the reduced coefficient of heat transfer from a fin decreases with increasing fin height and reducing fin pitch (i.e., with increasing φ); he therefore uses φ as a general parameter defining the finning of the tube. For a tube with circumferential fins this coefficient can be expressed in terms of the other fin parameters thus:

$$\varphi = \frac{u}{s} \left[2\frac{h}{u}\left(\frac{h}{d}+1\right) + \frac{\delta}{u}\left(1+\frac{2h}{d}\right) + 1 \right] \qquad (3.30)$$

Schmidt found by analysis of the effect of these parameters in heat transfer that h/s is the ratio which had the greatest influence, whereas h/d had less effect. The least significant effect is exerted by the ratio δ/u.

According to certain considerations, the optimum distance between fins should be at least twice the boundary-layer thickness δ [40]. For a laminar flow this quantity is obtained according to Blasius:

$$\delta = 5.83 \sqrt{\frac{\nu D}{w}} \qquad (3.31)$$

whereas for turbulent flow one may use the von Karman equation

$$\delta = 0{,}37\, D^{4/5} \left(\frac{\nu}{w}\right)^{1/4} \qquad (3.32)$$

It follows from the above that the optimum distance between fins in laminar and turbulent flows is not the same.

64 HEAT TRANSFER OF FINNED TUBE BUNDLES IN CROSSFLOW

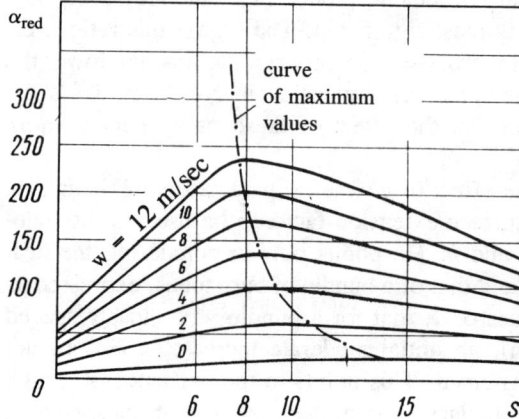

Figure 3.8 Heat transfer coefficient from a fin as a function of the pitch of the latter at different flow velocities.

Interesting experimental results concerning the above were obtained by Wegener [76] with straight, trapezoidally shaped fins with $h = 20$ mm. He reduced experimental data to obtain a relationship between α_{red}, the fin pitch, and the flow velocity. It is seen from Fig. 3.8 that at each flow velocity, there is a value of fin spacing s at which the heat transfer coefficient is a maximum. The fin pitch for maximum α_{red} increases with reduction in flow velocity. This also proves that the main role in this complex heat transfer mechanism is played by aerodynamic conditions, which basically govern the boundary layer thickness. Wegener used Eq. (3.32) for determining the fin spacing s for maximum coefficient (though without allowance for hydraulic drag); this spacing was found to be only 12% greater than the boundary layer thickness.

3. The Reference Velocity

The selection of flow velocity is of great importance for determining the similitude criteria for bodies in crossflow. Thus, the free-stream velocity w_0 is sometimes used as the reference velocity for flow over bare tubes (Fig. 3.9). The use of this velocity is valid only in the case when the tube diameter is small com-

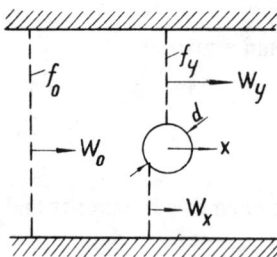

Figure 3.9 Selection of the reference velocity in flow over a tube.

pared with other channel dimensions, i.e., when the difference between velocities w_0 and w_y is small. When this difference is significant, the heat transfer from a finned tube is significantly affected by the velocity in the narrowest flow cross section, and the use of velocity w_0 for characterizing the mean heat transfer becomes inadequate. The most correct result in this case is obtained by using some mean flow velocity w_m, taking place in some location between f_0 and f_y. The mean value of this velocity can be integral, arithmetic or geometric, i.e.,:

$$w_{mi} = \frac{1}{d} \int_{-1/2 d}^{1/2 d} w_x \, dx \qquad (3.33)$$

$$w_{ma} = \frac{w_0}{2} \left(1 + \frac{w_y}{w_0}\right) \qquad (3.34)$$

$$w_{mr} = \sqrt{w_0 w_y} \qquad (3.35)$$

Schmidt [37] compared these three averaging methods and found that the difference between the mean integral velocity w_{mi}, which is a quantity rather difficult to determine, and the arithmetic mean w_{ma}, in tube configuration geometries actually encountered in practice does not exceed 14%, for which reason he recommends that w_{ma}, which can be easily determined, be used as the reference velocity.

It should be noted that the heat transfer coefficient for a finned tube will, in the final analysis, be affected by the flow velocity in the space between the fins. Schmidt analyzed this problem in detail [37] and determined the velocity distribution in the space between the fins, i.e., he determined the velocities w_{fin} at the fin tip and w_1 at its base. He found that these velocities are a function of the flow mode and of the finning parameters. The distribution of these velocities, written as w_{fin}/w_m or w_1/w_m was expressed in the form:

$$\frac{w_1}{w_m} \quad \text{or} \quad \frac{w_{fin}}{w_m} = f \left[\frac{(1 - R_0/\text{Re})^{R_1} (h/u)^{R_2} (d/h)^{R_3}}{(u/s)^{R_4}} \right] \qquad (3.36)$$

where R_0 through R_4 are constants.

The values of these constants were determined by reduction of data obtained in aerodynamic testing of a channel with a single finned tube. It was found that for Re > 10^4:

$$\frac{w_1}{w_m} = \frac{1 - (0.31/\text{Re}^{0.04}) (h/u)^{0.5} (d/h)^{0.25}}{u/s} \qquad (3.37)$$

$$\frac{w_{fin}}{w_m} = \frac{1 - (0.36/\text{Re}^{0.04}) (h/u)^{0.5} (d/h)^{0.25}}{u/s} \qquad (3.38)$$

whereas for $Re > 10^4$, respectively:

$$\frac{w_1}{w_M} = \frac{1-(2.16/Re^{0.25})(h/u)^{0.5}(d/h)^{0.25}}{u/s} \qquad (3.39)$$

$$\frac{w_{fin}}{w_M} = \frac{1-(2.50/Re^{0.25})(h/u)^{0.5}(d/h)^{0.25}}{u/s} \qquad (3.40)$$

The reference dimension for determining Re in Eqs. (3.37)–(3.40) is the streamline length l'.

It is seen from the above expressions that the velocity at the fin base is, naturally, lower than that at the tip. True, the difference is not too high, as can be seen from the values of R_0, which differ by 20% for the respective cases. Hence, in practical calculations this difference can be neglected without particular error, which allows one to use w_{fin} as the reference velocity. The ratio w_{fin}/w_m, is somewhat higher than 1, but does not exceed 1.25.

Rather satisfactory results are obtained when experimental data are reduced using Eqs. (3.37) and (3.39) with w_{fin} as the reference velocity (see Fig. 3.10). Use of this velocity as reference made it possible to correlate heat transfer from finned tubes over a wide range of finning parameters, represented in Fig. 3.7, with a high degree of accuracy. Note that the reduction in the velocity in the channel between the fins is proportional to the factor r_0/Re^{R1}, representing the friction drag. According to Blasius [75], the reduction in average velocity, relative to the centerline velocity, in turbulent flow in a channel with smooth walls, is proportional to $0.316/Re^{0.25}$. In channels with rough walls the reduction in velocity increasingly deviates from the Blasius law with increasing Re and becomes constant above a given value of Re. Equations (3.37)–(3.40) follow the same pattern: at $Re < 10^4$ the reduction in velocity is proportional to

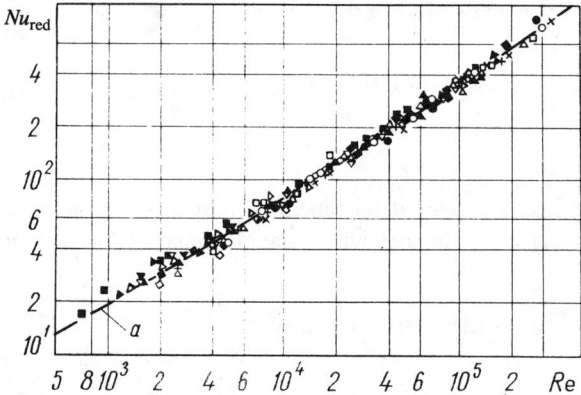

Figure 3.10 Reduction of experimental data for single finned tubes using velocity w_{fin} between fins as the reference velocity. (For legend see Fig. 3.7). The reference dimension is the streamline length l'.

$Re^{0.25}$, which corresponds to the Blasius law. At higher Re the exponent decreases (it was found to be 0.04).

It should be noted that the velocities w_m and w_{fin} have not come into extensive use in correlation of heat transfer from bundles of finned tubes. Usually, the velocity in the narrowest cross section of the bundle along the longitudinal or diagonal pitch (if the latter is smaller) is used as the reference velocity.

4. Effect of the Number of Rows in a Bundle on Heat Transfer

Heat transfer from a finned bundle depends directly also on the number of longitudinal rows of which the bundle is made up, since the heat transfer behavior of the first rows differs significantly from that of deeper-lying rows. Schmidt [37] evaluated constants m and c (Eq. 1.1) from experimental data for staggered and in-line bundles and as a function of the number of rows; results are presented graphically in Fig. 3.11. It is seen that the variations of c and m with a number of tube rows are different for in-line and staggered bundles respectively, particularly for $z < 3$. For both bundle configurations, the values of m increase with row number; those for the in-line configuration are somewhat higher than those with the staggered configuration. The ratio c_z/c_1 changes significantly for in-line bundles. A particularly steep downward jump is observed for the second row of the in-line bundle, which is followed by a gradual rise for the following rows. When the value of c_z is extrapolated to an infinite number of rows (c_∞) it is found that $c_\infty/c_1 = 1.35$ for an in-line and 0.95 for a staggered bundle.

Yudin, et al., [38] also found that the heat transfer coefficient for the first and last row of staggered bundle differed from that of the rows situated in between. They suggest that this be estimated by a coefficient k_z, equal to the ratio of the reduced heat transfer coefficient of row z to the analogous heat transfer coefficient of an internal row. In calculating the heat transfer of the zth row the reduced heat transfer coefficient of this row should be multiplied by the correction factor taken from Fig. 3.12. The corresponding graph for in-line bundles is given in Fig. 3.13.

Note that the heat transfer of the front rows depends strongly on the free-stream turbulence—increasing the latter significantly augments heat transfer from the first rows. According to Lapid and Schurig [41], placing meshes at the inlet to the bundle increased the heat transfer rate from the front rows by 40%. Unfortunately, flow turbulence was not measured in the studies reviewed above, which served as a basis for obtaining the correction factors derived by Lapin and Schurig. It can be assumed that it was not too high, and hence the suggested correction factors should be regarded as conditional.

68 HEAT TRANSFER OF FINNED TUBE BUNDLES IN CROSSFLOW

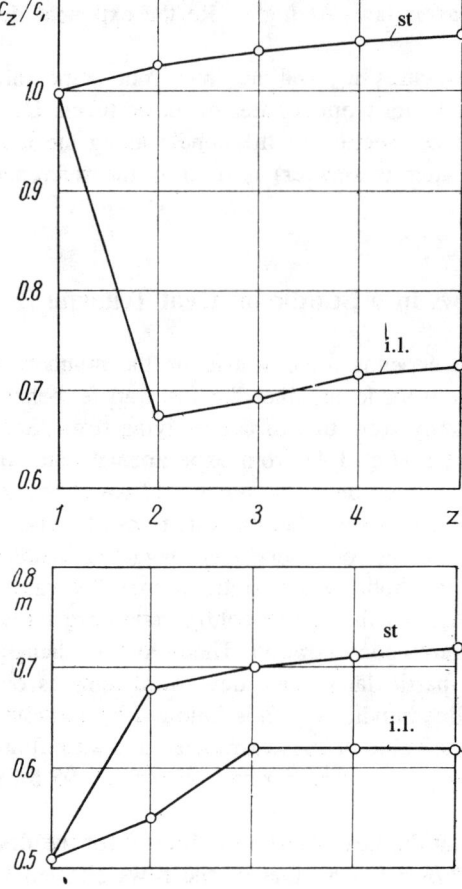

Figure 3.11 Effect of the number of longitudinal rows in a bundle of finned tubes on the value of exponent m and on the ratio c_z/c_1 (c_z is the constant for the zth row c_1, for the first row), i.l. pertains to the in-line and st to the staggered bundles configuration.

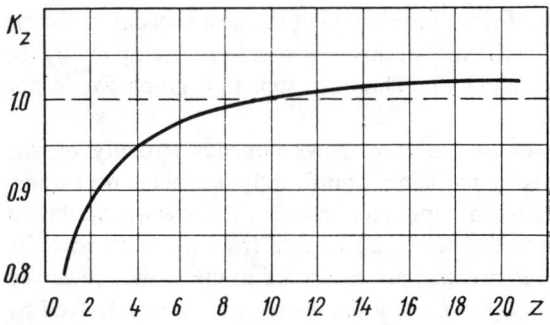

Figure 3.12 Correction factor k_z for the number of longitudinal rows in staggered bundles of tubes with circumferential fins.

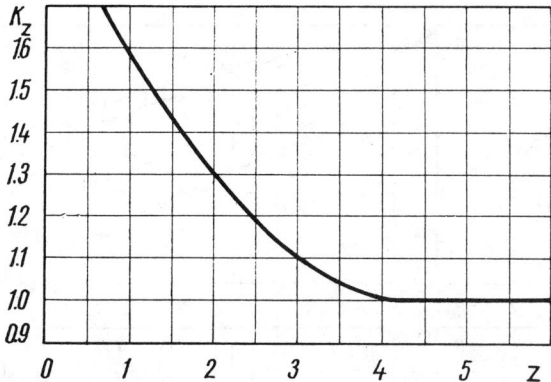

Figure 3.13 Correction factor k_z for the number of longitudinal rows in in-line bundles of tubes with circumferential fins.

3.4. EFFECT OF FINNED TUBE AND COOLANT THERMAL CONDUCTIVITIES ON HEAT TRANSFER FROM FINNED TUBE BUNDLES

Antuf'yev [8] was one of the first who attempted to investigate the effect of the thermal conductivities of the coolant and tube metal on heat transfer. However, for a long time, it was unclear how these factors can be used in design equations used for different coolants (with thermal conductivity λ_f) and different tube metals (with thermal conductivity λ_w).

Fortescue and Hall [26] performed a special study of this problem, and present experimental data obtained with a finned tube with $d \times s \times h = 29 \times 4.8 \times 28$ mm. The experiments were performed as follows: (1) single tubes of the same dimensions, but made of different metals (i.e., with λ_w varying, were placed in a test channel with air coolant); (2) a given tube was cooled by different fluids, (i.e., with λ_f varying). The reference dimension in data reduction was the equivalent diameter d_{eq}. The processed data are shown in Fig. 3.14, plotted in the form $St_{red} = f(Re)$. It will be seen that the experimental can be described by a single curve, given by the expression $St_{red} = f(Re\ \lambda_f/\lambda_w)$.

Analogous experiments were performed by Yudin with Tokhtarova [77]. They investigated heat transfer from five staggered bundles of tubes with $d \times s \times h = 23 \times 5 \times 10$ mm with $s_1/d = s_2/d = 2$, and also constructed of different materials (copper, aluminum, magnesium, carbon and stainless steel). They obtained a correlation in the form:

$$Nu_{red} = 0.079\,(\lambda_f/\lambda_w)^{-0.14}\,Re^{0.6} \qquad (3.41)$$

The results of testing a staggered bundle of stainless steel tubes with $d \times s \times h = 23 \times 2.5 \times 10$ with $s_1/d = 3$ and $s_2/d = 1.6$ cooled by air, carbon

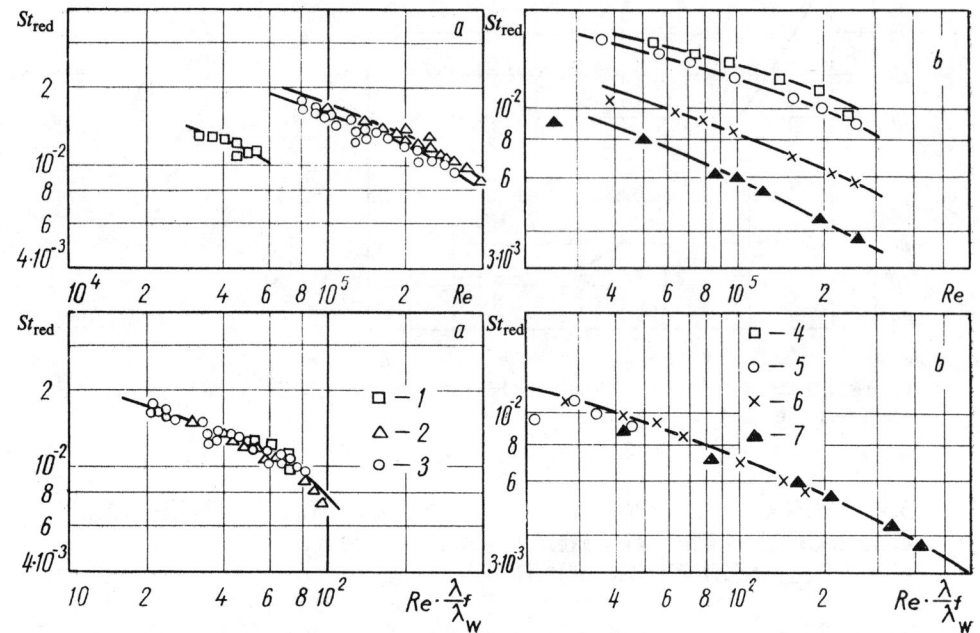

Figure 3.14 Effect of thermal conductivity of tube material and coolant on heat transfer from a finned tube. *a)* Data at λ_f = var: *1)* carbon dioxide; *2)* nitrogen; *3)* helium; *b)* data at λ_w = var: *4)* copper; *5)* duraluminum; *6)* carbon steel; *7)* stainless steel.

dioxide, argon and helium were correlated by the expression

$$\mathrm{Nu}_{red} = 0.4 \, (\lambda_f/\lambda_w)^{-0.3} \, \mathrm{Pr}^{0.4} \, \mathrm{Re}^{0.57} \qquad (3.42)$$

In addition, Yudin and Tokhtarova reduced their own data on the basis of the convective heat transfer coefficient given by Eq. (6.2), and obtained the following relationships:
For bundles with different fin materials at Re ≤ 3 · 10⁴

$$\mathrm{Nu} = 0.21 \, \mathrm{Re}^{0.65} \qquad (3.43)$$

For a bundle cooled by different fluids at Re from 4 · 10³ to 2 · 10⁵

$$\mathrm{Nu} = 0.048 \, \mathrm{Re}^{0.8} \, \mathrm{Pr}^{0.4} \qquad (3.44)$$

and at Re > 2 · 10⁵

$$\mathrm{Nu} = 0.013 \, \mathrm{Re}^{1.1} \, \mathrm{Pr}^{0.4} \qquad (3.45)$$

The reference dimension here is the tube diameter d. It can be concluded from the above that, although both these methods of experimental data reduction reflect the effect of thermal conductivity of the fin metal and of the coolant on heat transfer with sufficient precision, the second method, in which the data are reduced on the basis of the convective heat transfer coefficient is more general. In the latter case, the ratio λ_f/λ_w is eliminated, since by using the convective heat transfer coefficient, the differences in thermal conductivity are incorporated into the fin efficiency.

3.5. OTHER FORMS OF TEST DATA REDUCTION

The aforementioned material is based on data reduction using Eq. (1.1), as originated by Koschinke [78] and Grimison [79], and now in extensive use. A mention should be made, however, also of other forms of presentation of heat transfer from finned tube bundles.

A tube bundle consists basically of single tubes, arranged in a certain pattern, and naturally, there exists a relationship between heat transfer from a bundle and from a single tube. When the distance between tubes within a bundle is sufficiently large, the heat transfer from a single tube under both conditions should be identical. If the streamline length l' is used as reference dimension in Nu and Re, which makes sense from the physical point of view, then, according to Krischer it becomes possible to describe the heat transfer from different bodies: cylinders, spheres, plates, prisms, etc., placed in transverse coolant flow by a single curve [81]. On the other hand, a bundle of tubes can also be imagined as a system of channels between separate bodies, over which a coolant flows. The heat transfer in such channels is determined from the familiar functional relationship of Hausen [80], written for the general case in the form

$$\mathrm{Nu}_{d^*} = f\left(\mathrm{Pe}_{d^*};\ \frac{d^*}{l'};\ \mathrm{Pr}\right) \qquad (3.46)$$

where d^*/l' is the equivalent channel diameter

$$\frac{d^*}{l'} = \frac{d}{l}$$

in the case of flow in tubes, $4f/F$ for channels of any cross section;

$$\frac{4f_0\psi}{F}$$

for channels with periodic blocking of the cross section in the direction of flow, including tube bundles in transverse flow. In the latter, f_0 is the unobstructed

72 HEAT TRANSFER OF FINNED TUBE BUNDLES IN CROSSFLOW

cross section of the channel and ψ is the channel porosity index (fraction of free space in the obstructed channel).

When reducing data according to Eq. (3.46), the reference dimension in Nu_{d*}, and Re_{d*} and Pe_{d*} is $d*/l'$. The heat transfer coefficient of the body, which is used in Nu_{d*}, is referred to as the difference between the free-stream temperature and the mean temperature of the body's surface.

Figure 3.15 shows a graph describing the heat transfer in different channels with axial flow, and also heat transfer from various bodies in crossflow, including bundles of finned tubes [39]. The experimental data were reduced with Eq. (3.46). This form of data reduction, capable of correlating data for such a wide range of body shapes, definitely deserves attention.

As is seen from Fig. 3.15, turbulent heat transfer from different bodies (including finned and bare surfaces) placed in different channels differs little when data is reduced in this manner (curves b through h). Data on heat transfer from bodies placed in a flow, correlated by curve i, corresponds to the thermal and hydrodynamic inlet region of a channel with a laminar boundary layer. The

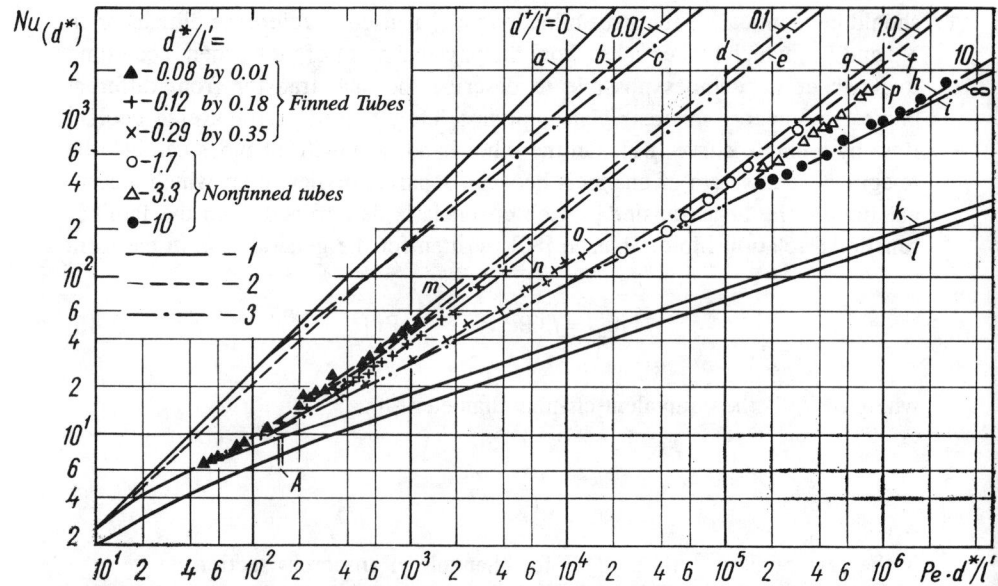

Figure 3.15 Correlations of heat transfer from various surfaces a) Thermal inlet region of channel: g) and h) thermally and hydrodynamically developed turbulent flow in channels and flow over bodies at different values of reduced diameter $d*/l'$; i) inlet region of body in external flow with laminar boundary layer; k) and l) inlet region of hydrodynamically developed laminar flow in a flat channel and in a tube; m) through p) finned and nonfinned tubes; 1) channels with laminar flow; 2) channels with turbulent flow; 3) bodies in external flow.

segment of the graph between curves i and l pertains to laminar flow and corresponds to transition from an inlet region with laminar flow is possible only up to Re = 2300 and that a minimum length of ten diameters is needed for the formation of a laminar velocity profile in the inlet region, it was found that

$$\text{Pe}\frac{d^*}{l'} = 166$$

which is represented in the graph by point A. Note that the data obtained for bundles of finned and bare tubes at high Re_{d*} lie somewhat about curve i.

The above graph makes it possible to compare the heat transfer from finned surfaces under different conditions of flow over them. If it is assumed that at moderate Re the conditions of flow between the fins will be laminar, or will correspond to those in an inlet region of a channel with a laminar boundary layer, then the values of Nu_{d*} corresponding to this lie 20% lower as compared with, for example, the hydrodynamic inlet region of a channel.

A few words now on work by Kast [44]. He investigated heat transfer in beds formed by bodies of various shapes such as spheres, prisms, cylinders, etc., and suggested that heat transfer in bundles of finned tubes also be investigated on the basis of a plot of

$$\alpha_b/\alpha' = f(d^*/l') \tag{3.47}$$

where α' is the coefficient of heat transfer in a single finned tube in crossflow, where as α_b is the same for a tube in a bundle, while d^*/l' is given by Eq. (3.46).

Kast treats the ratio α_b/α' as a kind of bundle geometry factor. The bundle is defined as a random ensemble of tubes. At high s_2/d, and respectively d^*/l' each body within the bed can be regarded as single, and hence $\alpha_b/\alpha' = 1$. For closely-spaced tubes, i.e., at low d^*/l', the flow can actually occur without separation and without vorticity formation, as in very narrow and long channels. Hence ratio α_b/α' is also equal to 1. Between these cases there exists a region within which the bodies, arranged one after the other, are situated in a vortex wake. Figure 3.16 shows graphs of this kind, constructed from experimental data for staggered and in-line bundles of finned tubes, obtained by a number of investigators. The data used were obtained at $\text{Re}_{l'} = 10^4$. It is seen that for both bundle configurations ratio α_b/α' has a certain maximum depending on the bundle geometry (relative pitch b). The curves for both kinds of bundle geometry for b between 3 and 4 are identical. Then at $b = 2$ the difference between these configurations becomes significant. The limiting value of α_b/α' of the densest in-line arrangement is 0.5, being 1 for the staggered geometry. This difference is attributed to the fact that, for the in-line arrangement, only one half of the heated surface of the tubes is involved in heat

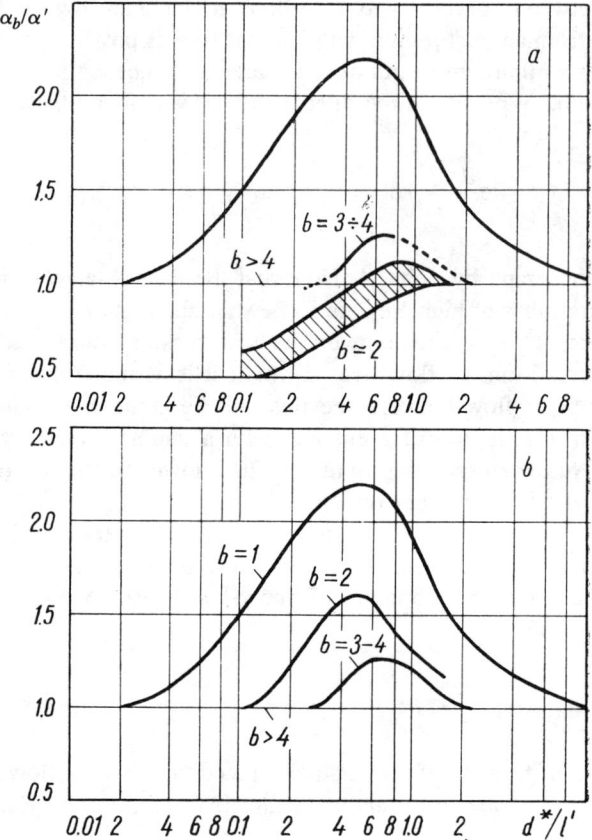

Figure 3.16 Transfer of heat from a bundle of finned tubes. *a)* in-line configuration; *b)* staggered.

transfer, whereas the other half is shaded. At $b > 4$ we have $\alpha_b/\alpha' = 1$ for both configurations.

These graphs can be used for following the effect of the number of longitudinal rows on heat transfer. For a single row $\alpha_b/\alpha' = 1$. As z is increased to 3–4 the value of α_z at $\alpha_b/\alpha' > 1$ increases, while decreasing at $\alpha_b/\alpha' < 1$.

Unfortunately, this method of data presentation makes it difficult to explain the effect of other finning parameters, such as h/u and h/d. It is also difficult to represent the relationship with bare-tube bundles—the value of d^*/l' in the latter lies within the limits of 0.6 / d^*/l' 4, i.e., in the region where α_b/α' decreases as a function of d^*/l'. It is likely that these problems will be investigated further in the future.

CHAPTER
FOUR

EXPERIMENTAL TECHNIQUES

Our experimental studies of heat transfer and patterns of flow over bundles of finned tubes at Re up to 10^6 were performed with specially constructed experimental equipment and specially developed experimental techniques. The high Reynolds numbers were attained by operating at coolant pressures of up to 25 bar gage.

The experiments performed under high pressures differ somewhat from those performed under atmospheric pressure. Allowance for this was made in designing the test-stand components, in selecting monitoring instrumentation, and also in designing the experimental procedures. We were guided by reference to the work described by Brauer [17] and Jameson [33] and by the recommendations of Krischer and Loos [81].

The shape of the fins studied was selected taking into account manufacturing considerations. The actual fabrication of finned tubes is a rather complicated matter, and therefore fabrication technology is an important factor, at times controlling the economic feasibility of a given fin type. During the past decade a new technology has become available for producing seamless tubing with continuous spiral fins having a trapezoidal cross section [22]. The use of this advanced type of fins combined with an efficient and economical technology of their production makes it possible to use them extensively in a number of industries. This is the type of fin which was selected for our studies.

The principal focus of our work was obtaining experimental data which could be used to develop working equations for heat transfer and hydraulic drag in finned tube bundles at high Re.

The use of advanced methods of correlation of experimental data, based on the convective heat transfer coefficient, was made possible by special experiments in which the distribution of the heat transfer coefficient over the fin could be determined.

Since the rate of heat transfer depends to a large degree on the finning geometry and bundle configuration, a great deal of attention was given to determining the effect of these factors and obtaining their optimal values. We investigated bundles with different dimensions of the spiral fins, and also with different bundle configurations obtained by varying the relative pitches.

Due attention was given to the study of the mechanism of the flow over a finned tube. Special experiments were performed for determining local heat transfer features over a wide range of Re.

4.1. THE EXPERIMENTAL ARRANGEMENT

The experimental arrangement consisted of a closed aerodynamic loop (Fig. 4.1), designed and constructed for gas-coolant pressures up to 25 bar gage. Basically the test loop consists of the test section *(2)*, connecting piping *(1)*, an expansion chamber *(3)*, moisture and oil trap *(4)*, and a drum *(5)*, with the blower *(6)*, placed in it. Air, sucked in from the drum 5 (1.5 m^3) flows through trap *(4)* and expansion chamber *(3)* to the test section, from where it is returned

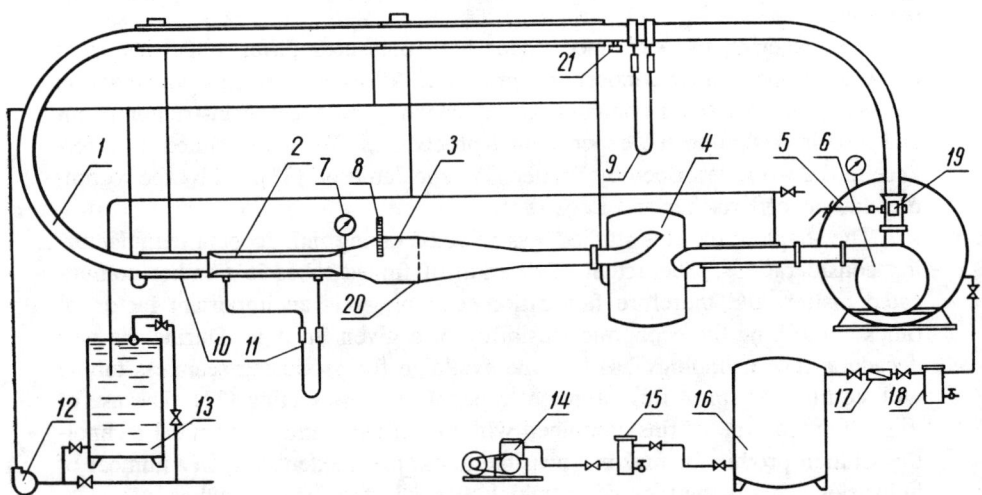

Figure 4.1 Schematic of experimental arrangement. *1)* Connecting piping; *2)* test section; *3)* expansion chamber; *4)* oil trap; *5)* drum; *6)* air blower; *7)* pressure gage; *8)* thermometer; *9)* differential manometer; *10)* water jacket; *11)* differential manometer; *12)* circulation pump; *13)* water tank; *14)* high-pressure compressor; *15)* moisture and oil trap; *16)* receiver; *17)* pressure reducer; *18)* linear moisture and oil trap; *19)* damper; *20)* grid; *21)* thermocouple insertion stub.

to drum *5* along connecting piping (200 mm in diameter). The pumped gas circulating within the loop is heated up due to the dissipation of kinetic energy. The air temperature is maintained at the desired level by a water look, consisting of water jacket *10,* circulation pump *12,* and water tank *13*. Water in the tank is maintained at constant temperature by replenishing from the mains by means of an automatic float-controlled valve, and hot water is discharged.

The experimental equipment has been specifically designed to obtain high Re in the test section. A wide range Re is obtained by adjusting the flowrate of air through the test section (by controlling the rpm of the air-fan motor, and also by throttling on the intake stub by damper *19),* or by changing the air pressure in the loop. The required air pressure level is achieved by pressurizing the loop using compressor *(14)* (40 bar, 60 Nm^3/hr). Air from the compressor is supplied through the moisture and oil trap *(15)* to a 3 m^3 capacity receiver *(16),* whence it flows to the *(5)* through pressure reducer *(17)* and linear moisture and oil trap *(18).*

The pressure in the receiver and in the experimental loop is maintained automatically within preset limits by means of the pressure reducer *(17)* which has a capacity of 20 Nm^3/min.

The air flowrate is determined from the pressure drop across a standard orifice, measured by a water-filled differential manometer *(9).* For low air flowrate, the differential manometer was replaced by a micromanometer. The standard orifices were designed and constructed in accordance with Preobrazhenskiy's [85] specifications with 165 and 65 mm diameter holes. The orifices were calibrated against the Pitot-Prandtl tubes. The air flowrate measured by the orifice was used for calculating the air velocity in the test section. The inlet to the test section was smooth in order to obtain a uniform velocity distribution. For the same purpose a grid *(21)* was placed in the expansion chamber. The velocity field immediately ahead of the bundle was checked by the Pitot-Prandtl tubes. The horizontal and vertical velocity profiles in the test section zone were equalized and made fully symmetrical by adjustment of the grid.

The flow temperature was measured in two places: ahead of the orifice and ahead of the test section. The temperature ahead of the orifice was measured by a resistance thermometer placed in the channel, while that ahead of the test section was determined by mercury thermometer *(8)* with 0.1 °C graduations, placed in the expansion chamber.

Heat leakage was eliminated by properly insulating the expansion chamber and the manifold ahead of the test section.

The Test Section

The test section and the expansion chamber are of rectangular cross section. The test section is 1200 mm long and 200 mm wide, the height being varied depending on the transverse tube pitch by wooden plates. The interior part of the test section, consisting of the tube bundle being tested, was retractable. It is fastened to the side wall, forming the lid of this section. To reduce thermal losses, the interior part of the walls was lined with plates of wood and of plastic insulating material (see Fig. 4.2), so that the air flowing through the test section would not come into contact with the metal of the walls.

The cross section (350 × 350 mm) of the expansion chamber was selected so as to obtain an optimum ratio between the cross sectional area of the expansion chamber and the test section, taking account of the feasibility of construction.

The experimental studies were performed in staggered bundles with different finned-tube dimensions and different longitudinal and transverse pitches. Twenty-one bundle configurations were used. The studies were performed with two main objectives: 1) to investigate the effect of finning geometry; 2) to determine the effect of tube location within a bundle. Data on the bundle geometries used are given in Tables 1 and 2.

For the first objective, it was necessary to minimize the effect of tube location within the bundle. For this reason the bundles under study were configured in such a manner that the ratios of flow cross sections (i.e., $f_1/2f_2'$) in the transverse and diagonal sections with pitches s_1 and s_2', respectively (Fig.1.2), would be as close as possible in experiments with different fin geometries. This, in effect, provided equivalence of flow conditions over the different bundles. For technical reasons we could not rigorously maintain this condition, and slight deviations were permitted. The bundles were assembled from machined aluminum 'dummy' (i.e., unheated) tubes, plus tubes on which the measurements were made. the latter being machined from steel. It is seen in Fig. 4.2 that the top and bottom walls of the test section intersect the bundle along tube axes, i.e., tube halves were fastened to these walls in order to obtain an "infinite bundle." The experiments were performed with seven-row bundles. A calorimeter tube was placed in the first and fifth rows in the direction of flow. It is known from several of the previously mentioned studies that heat transfer stabilizes at the third or fourth longitudinal row. This was confirmed by our preliminary experiments in which the calorimeter tube was placed in different rows.

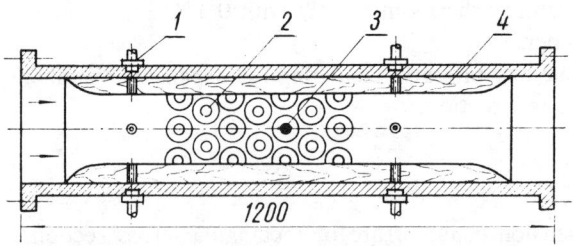

Figure 4.2 *1)* Pressure drop measurement station; *2)* dummy finned tube; *3)* calorimeter tube; *4)* wooden plate.

TABLE 1 Description of Bundles

Bundle number	Fin dimensions			Bundle pitches		Relative pitches		
	d, MM	s, MM	h, MM	s_1, MM	s_2, MM	s_1/d	s_2/d	$f_1/2f_2^1$
1	32	4	4	70.4	41.6	2.2	1.3	0.92
2	32	4	6	70.4	41.6	2.2	1.3	0.94
3	32	4	9	72.6	43.7	2.27	1.36	0.92
4	32	4	13.5	76.0	46.0	2.38	1.46	0.94
5	32	6	6	70.4	41.6	2.2	1.3	0.92
6	32	6	9	70.4	61.6	2.2	1.3	0.94
7	32	6	13.5	76.0	46.6	2.38	1.46	0.87
8	32	8	9	69.3	40.6	2.17	1.27	0.94
9	23	6	6	61.4	33.6	2.67	1.46	0.91
10	23	6	9	61.4	33.6	2.67	1.46	0.94
11	23	6	13.5	63.0	38.0	2.97	1.65	0.88
12	23	4	4	61.4	33.6	2.67	1.46	0.91

Fin thickness in all configurations $\delta_2/\delta_1 = 2/1$ mm, fin angle of attack from 2 to 4°.

TABLE 2 Description of Bundles

Bundle number	Fin dimensions			Bundle pitches		Relative pitches	
	d, MM	s, MM	h, MM	s_1, MM	s_2, MM	s_1/d	s_2/d
13	23	6.5	10	72	41	3.13	1.78
14	23	6.5	10	93	41	4.04	1.78
15	23	6.5	10	93	45.6	4.04	1.98
16	23	6.5	10	63	48.6	2.74	2.11
17	23	6.5	10	63	37.7	2.74	1.64
18	23	6.5	10	94.5	49.2	4.11	2.14
19	23	6.5	10	86	46.2	3.74	2.01
20	23	6.5	10	61.4	33.6	2.67	1.46
21	23	6.5	10	95.0	33.6	4.13	1.46

Fin thickness $\delta_2/\delta_1 = 2/1$ mm, fin angle of attack from 2 to 4°.

4.2. DETERMINATION OF HEAT TRANSFER FROM A BUNDLE

1. Selection of a Thermal Simulation Technique

Two methods are available for determining the heat transfer from a tube within a bundle—these are the local and complete thermal simulation methods. In the first case only the measuring tube—the calorimeter tube—is heated, whereas the remaining tubes within the bundle remain cold. The mean arithmetic flow temperature entering and leaving the bundle or the free-stream temperature is then taken as the temperature of the flow, which, in the case of moderate temperature differences between the calorimeter tube wall and the flow, are almost equivalent.

Under the second method, all the tubes are heated and thermocouples are placed on the surface of tubes in all the rows (on one tube of each row). The flow temperature is then defined as the arithmetic mean temperature of air entering and leaving the bundle, whereas the temperature of the surface of a tube in an inner row is obtained as the arithmetic mean temperature of the surface of all the measuring tubes within the bundle. Evidently, in this case it is more difficult to determine exactly the flow temperature ahead of each calorimeter tube as is done, for example, in the local method. The accuracy with which the heat transfer coefficient is determined using this method is lower than when using the local method.

The local simulation method is very convenient in investigating heat transfer from a bundle of tubes, particularly in the case of bare tubes. The question arises, however, as to how valid this method is in producing data applicable to the case of heat exchangers with finned bundles.

Similitude theory requires that processes reproduce by the model be similar to those occurring in real heat exchangers. For this it is necessary to provide for geometric similarity and similarity of boundary conditions.

The propagation of heat within a moving coolant is described by the differential energy equation, according to which the temperature field in a flow is a function of the velocity components and of thermal conduction and mixing.

This means that, in order to provide for similar conditions, it is necessary to attain both hydrodynamic and thermal similitude. As was previously mentioned, the flow in finned tube bundles stabilizes after the second or third row, which implies that the flow pattern obtained inside a bundle of finned tubes is independent of boundary conditions and is fully defined by the configuration of the bundle proper and by the finning geometry. Accordingly, the first of the two conditions for similarity of processes can be easily satisfied by assembling a bundle geometrically similar to the actual heat exchanger. However, attainment of thermal similitude is a more difficult problem.

The use of the local simulation method involves departure from rigorous similitude, because heat transfer in all the tubes within the model bundle does not occur simultaneously, with the result that the temperature fields of the model, and consequently, the physical properties of the coolant, do not correspond to temperature fields in the real heat exchangers. In addition, in this method, the flow onto the calorimeter tube under study is at the same temperature in all the points. This may not be the case where the whole bundle is heated, and where streamwise variations of temperature may be introduced as a result of heat transfer from the upstream tubes. The heat transfer that occurs then is similar to that in a real heat exchanger only in the case when the temperature in the entire flow onto the given row is the same, i.e., if the mainstream and the boundary layer separating from upstream rows tube are well mixed. The degree of this mixing depends on the bundle shape and on Re.

Experience accumulated by our institute in this field [32] shows that the local method is satisfactory for studies with bare-tube bundles. Furthermore, Antuf'yev [8] and Yudin and Tokhtarova [86] recommend that the method may be usually used for finned bundles, though the latter authors note that, for closely pitched tubes and for flow in the subcritical range (Re $< 5 \cdot 10^4$), the heat transfer coefficient determined by the local technique has a higher value than that determined by the complete simulation method.

Given the fact that the local simulation method allows a more exact determination of the heat transfer coefficient of a tube, and also that it was planned to perform the experiments also at Reynolds numbers above the critical (Re $> 10^5$), we selected local simulation as the experimental method. Subsequently, this selection was justified in the study by Yudin and Takhtarova [87]. Figure 4.3 is a plot of their results for the ratio of Nusselt numbers obtained by the complete and local simulation techniques (Nu_{comp}/Nu_l), as a function of the longitudinal pitch s_2/d and the value of Re. It is seen that the heat transfer coefficient from staggered finned bundles, obtained by the local thermal simulation technique are, at Re $< 4 \cdot 10^4$, higher than the same coefficients obtained by the complete simulation method. At $s_1 = s_2 = 2$ this difference is $\sim 6\%$, rising to $\sim 11\%$ at $s_1 = 3d$ and $s_1 = 2d$. At constant transverse pitch ($s_1 = $ const), the difference in heat transfer coefficients obtained by the two methods decreases with increasing longitudinal pitch (s_2). This difference also decreases significantly with increasing Re.

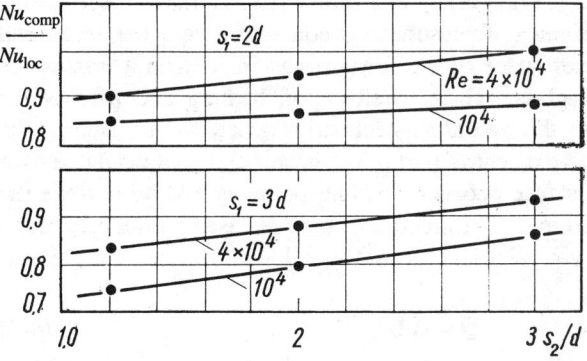

Figure 4.3 Comparison of heat transfer measured in staggered finned tube bundles by the local and complete thermal simulation techniques.

2. The Mean Heat Transfer was determined for steady coolant velocity and heat input on the basis of the reduced heat transfer coefficient, which was determined from the expression

$$\alpha_{\text{red}} = \frac{Q}{F(t_1 - t_f)} \qquad (4.1)$$

where F is the total finned surface, t_1 is the temperature of the outside surface of the tube at the fin base and t_f is the flow temperature.

The quantity of heat transmitted was determined by monitoring the electrical power dissipated in the calorimeter tubes and the flow temperature t_f was determined from a mercury thermometer, placed in the flow stabilization region. It was checked against readings of thermocouples placed in the dummy tubes. The temperature t_1 was determined by means of thermocouples embedded into the surface of the calorimeter tube at the fin base.

3. Local Heat Transfer was investigated using a special tube with a thin heater strip placed on the fin; the heat flux density q_i was determined over the surface of the strip. The local heat transfer coefficient was calculated from the equation

$$\alpha_i = \frac{q_i}{t_i - t_f} \qquad (4.2)$$

where local temperature t_i was measured by thermocouples. The quantity q_i of heat released on a segment of the heating element was calculated in direct proportion of its length, since its thickness was constant.

4. Calorimeter for Determining the Mean Heat Transfer

The design of the monitoring tube and electric heater is shown in Fig. 4.4. A copper sleeve is used to equalize the heat flux transmitted by the heater to the steel finned tube. The heater *(4)* consists of a thin-walled stainless-steel sleeve, insulated by a thin layer of mica. To ensure tight contact between the heater and the copper sleeve, the inner space of the former was filled with a mixture of firebrick clay and winter-glass, which swells upon heating and presses the heater to the sleeve. Two thin stainless-steel strips, used for measuring the voltage drop, are welded to the central part of the heater and well insulated over their entire length. The distance between welding points is 100 mm, while the overall heater length is 200 mm. The amount of heat released by the tube along a 100 mm long segment was calculated from the voltage drop and the current:

$$Q = \Delta U I \qquad (4.3)$$

EXPERIMENTAL TECHNIQUES 83

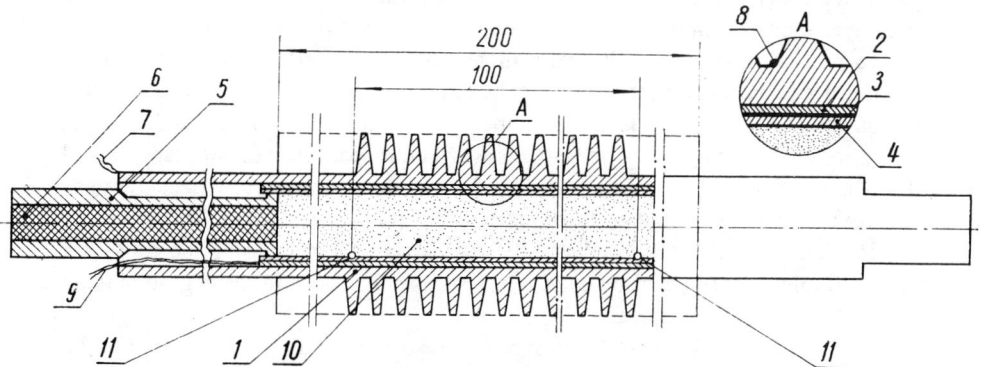

Figure 4.4 Measuring calorimeter tube with electric heaters. *1)* Finned tube; *2)* copper sleeve; *3)* mica; *4)* heating element; *5)* current bushbars; *6)* asbestos; *7)* resistance-thermometer (or thermocouple) wires; *8)* thermocouple beads; *9)* voltage-drop measurement wires; *10)* filler; *11)* voltage-drop measurement contacts.

Voltage drop ΔU was measured by a potentiometer, whereas current I was determined by a class 0.15 ammeter.

Heating was by direct current, supplied by selenium rectifiers fed from a three-phase power supply through voltage regulators.

The wall temperature of the calorimeter tube was measured by a resistance thermometer or by thermocouples. The resistance thermometer was placed in a helically-shaped groove at the fin base, the groove extending over an axial length of 100 mm at the mid-point of the tube span. The resistance thermometers were constructed of 0.05 mm diameter copper wire and the helical grooves were 0.10 mm deep; the resistance thermometers thoroughly electrically insulated from the calorimeter body. In certain cases the temperature was determined from readings of four thermocouples, placed within the wall as the mid-point of the tube span and at circumferential locations of 0, 75, 135 and 180°, measured from the stagnation point. The thermocouples consisted of lacquer- and silk-insulated, 0.15 and 0.07 mm diameter copper-constantan wires and were placed in 1 mm deep and 0.5 mm wide grooves. The latter were filled up from the top by a specially shaped copper strip, and then tube was then turned smooth. To double check, both resistance thermometers and thermocouples were used for measuring the temperature in some of the experiments. The agreement between the results was satisfactory.

The mean temperature of the total surface of the finned tube was determined by resistance thermometers, which were placed on the surface of the fin-bearing tube, on the sides and on the tip of the fin. The mean temperature of the finned tube was calculated in proportion to the surfaces under measurement.

The thermocouples and the resistance thermometers were calibrated in a constant-temperature oil bath against a standard mercury thermometer with

0.1 °C scale divisions. The thermocouple emf and the voltage drop across the resistance thermometers were measured by a PPTN-l potentiometer.

To reduce heat losses, the calorimeter tube is designed in such a manner that there can be no direct contact between it and the metal walls of the test section (Fig. 4.5). The calorimeter tube *(3)* is centered on inner walls *(1)* of plastic insulating material. The busbars were connected to current leads *(7 and 8)*. Asbestos filler *(5)* prevents leakage of air through imperfectly packed openings when working at above-atmospheric pressures.

To check for the magnitude of thermal losses from ends of the tube, six thermocouples were placed on each of the measuring tubes in the region outside the (central) measuring zone.

As a whole, the calorimeter design has justified itself. It made possible the determinations of the heat transfer coefficient without exceeding the error limit which was set at 6–8%.

5. Calorimeter for Measuring the Local Heat Transfer

The local heat transfer coefficients were measured by a specially constructed experimental tube with the following dimensions:

diameter of tube supporting the fins	$d = 32$ mm,
fin height	$h = 13.5$ mm,
fin thickness at its base	$\delta_2 = 2$ mm,
fin thickness at its tip	$\delta_1 = 1$ mm,
tube length	$l = 200$ mm
angle of attack of fin	2–4°.

The calorimeter was designed in such a manner that it would be possible to vary the fin pitch during the experiments from 4 to 7 mm. The calorimeter body was machined from steel. The heating element (Fig. 4.6) consisted of a thin (0.05

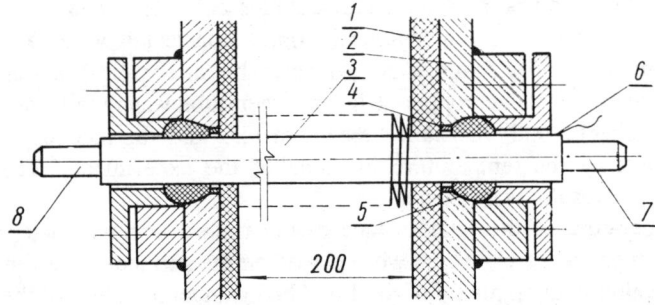

Figure 4.5 Placement of measuring tube in the test channel. *1)* Plastic insulating material; *2)* steel wall of channel; *3)* measuring tube; *4)* textolite sleeve; *5)* asbestos filler; *6)* thermocouple wires; *7)* and *8)* current leads.

EXPERIMENTAL TECHNIQUES 85

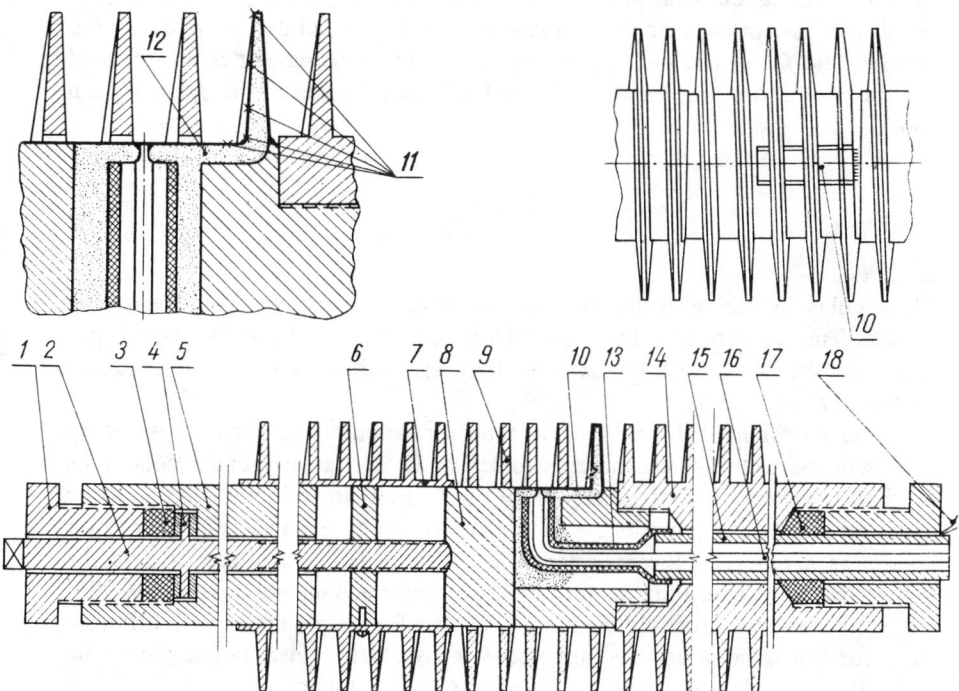

Figure 4.6 Measuring tube for investigating local heat transfer coefficient. *1)* Packing nut; *2)* bolt; *3)* packing; *4)* washer; *5)* housing; *6)* nut; *7)* sliding fins; *8)* fixed housing; *9)* variable-pitch fins; *10)* heating strip; *11)* thermocouple attachment points; *12)* filler; *13)* insulation; *14)* housing; *15)* insulation tube; *16)* busbar; *17)* packing; *18)* thermocouple and voltage-measurement wires.

mm thick) silicon-nickel tape which passed over the fin. The tape was heated by an alternating current. Five copper-constantan thermocouples of 0.07 mm diameter wire were welded to the tape. (The thermocouple locations are seen in the figure).

The local heat transfer coefficients measured by this calorimeter corresponded to conditions of constant heat flux on the heating plate (q = const). In addition to changing the fin pitch in the course of the experiments, it was possible to change the location of the heating plate by rotating the tube around its axis.

4.3. DETERMINATION OF FLOW VARIABLES

The air velocity used as the reference was calculated from the flow in the narrowest passage between two tubes on the basis of s_1 or s_2' and was determined by the assumption of conservation of mass. If f_y, w_y and ρ are respectively the flow-passage cross sectional area, velocity and density of air in the bundle, and f_1, w_1 and ρ_1, respectively, represent these quantities upstream of the measuring orifice, then the flow velocity is calculated from $f_1 w_1 \rho_1 = f_y w_y \rho$ thus

$$w_y = \frac{f_1 w_1 \rho_1}{t_y \rho} \qquad (4.4)$$

To calculate ρ_1 we measured the air temperature upstream of the measuring orifice. The pressure drop in the pipe from the bundle to this orifice was negligible, and hence ρ_1 was calculated on the basis of the pressure of air upstream of the orifice.

The maximum velocity used in the correlations does not always reflect precisely the nature of the process under study, particularly under conditions when this velocity differs greatly from the free-stream velocity (i.e., in more compact bundle configurations). As recommended by Schmidt [37], it is better to use as a reference the mean velocity between two tubes measured in the region between the leading point of the finned tube and the point with $\varphi = 90°$. To be able to very precisely compare heat transfer from bundles with different configurations, the mean integral velocity w_m was determined by integrating the velocities in the bundle cross section over a wide range of s_1/d.

The ratio of w_m to w_y is plotted in Fig. 4.7 as a function of s_1/d. It is seen from the figure that increasing s_1/d gives an increase in w_m/w_y results in a relative increase in the rate of heat transfer from the surface to the flow; however, w_m/w_y is not directly proportional to s_1/d. W_m/w_y increases more rapidly with s_1/d at a low s_1/d. Hence, in denser bundles, where $s_1/d < 2.5$, w_m is

Figure 4.7 Ratio of flow velocities w_m/w_y vs the relative transverse pitch of a finned bundle.

closer to the free stream velocity than in w_y, and the use in this case of w_y in characterizing heat transfer results in excessive error. In bundles where $s_1/d >$ 2.5 the mean velocity is closer to the maximum, and in this case the latter can be used without much error. In our experiments and correlations on the effect of bundle configuration on heat transfer, s_1/d was varied from 2.67 to 4.13, and hence the use of maximum velocity w_y as the reference is justified.

The total pressure drop over the bundle was determined as the difference between the static pressures in the test section upstream and downstream of the bundle respectively. The static-pressure drop was measured by a liquid-filled differential manometer (*11* in Fig. 4.1) and was calculated from the expression

$$\Delta p = (\rho_h - \rho') H \tag{4.5}$$

where ρ_h is the density of the manometer fluid, ρ' is the density of air above the manometer fluid, and H is the difference in the heights of the columns of the manometer fluid.

Three differential mamometers were used: two of these were mercury filled, whereas the third used water. At low pressure drops, the readings were taken from the water manometer and were checked by the mercury instrument; at high pressures, when the readings of the water manometer went off scale, the pressure drop was taken as the arithmic mean of the readings of the two mercury manometers. The differential manometers were connected, by way of brass tubes, to stubs, installed on three walls of the test section—upstream and downstream of the bundle. The stubs communicated with holes not more than 1 mm in diameter, drilled exactly perpendicular to the inner surface of the channel wall.

The static pressure of air upstream of the bundle was measured by a standard gage (0.2 precision class), or by a U-shaped water manometer, depending on the test conditions.

The hydraulic drag of the bundles was determined under steady flow conditions. The effect of the channel walls was not taken account of in correlating the data for pressure drop; the nondimensional correlation of the data was based on the Euler number (Eu).

4.4. REMARKS ON THE EXPERIMENTAL WORK

The actual experiments were preceded by preparatory studies, consisting of checking the velocity distribution in the inlet part of the test section, and also in checking the operation of the static-pressure sensor, airtightness of the test section, etc. To verify the quality of test equipment performance, the preliminary experimental points were compared with similar data of other investigators.

The Reynolds number was varied by changing the pressure and circulation rate in the loop. The entire range of Re to be explored during a given experiment was subdivided into individual subranges, which had a precisely specified pressure level corresponding to them. When a given flow condition was attained, the thermocouple and thermometer readings were taken. The differential manometers were also used at the same time. During these operations the specified pressure level, temperature and flow velocity, as well as the thermal load on the calorimeter heater, were maintained rigorously constant. To eliminate random errors, each experiment was repeated twice.

It should be noted that the temperature and velocity of the air did not vary greatly over the entire range of Re under study (the temperature ranged on the average from 30 to 70 °C, and the flow velocity from 25 to 35 m/sec). The temperature difference between the tube wall surface and the air was maintained within the limits from 25 to 70 °C. Calorimeter tubes were placed in the first and fifth rows of the bundle.

A separate series of experiments were performed to obtain local heat transfer coefficients and, hence, the coefficient of nonuniformity of heat transfer over the fin surface.

For data reduction information is required on the physical properties of the coolant: namely density ρ, specific heat c_p, the dynamic viscosity μ and thermal conductivity λ_f. These data were taken from the books by Petukhov [82] and Mikheyev [83].

The physical properties of air are temperature dependent and, with the exception of density and kinematic viscosity, depend little on pressure. The kinematic viscosity is given by the expression

$$\nu = \frac{\mu}{\rho} \tag{4.6}$$

An increase in pressure results in a significant reduction in kinematic viscosity and, hence, in an increase in Re for a given flow velocity. Figure 4.8 shows a nomogram, constructed from data presented by Vargaftik [84], which illustrates the variation in kinematic viscosity of air vs pressure and temperature.

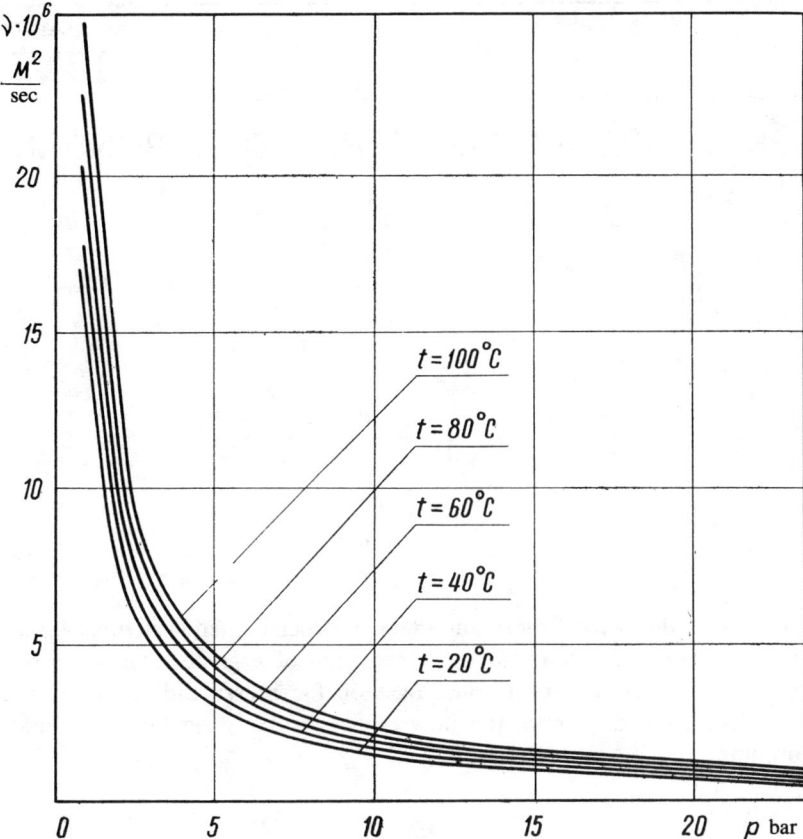

Figure 4.8 Nomogram of the kinematic viscosity of air vs pressure and temperature.

CHAPTER
FIVE

PRESSURE DROP IN FINNED TUBE BUNDLES

Pressure drop is the second most important parameter after heat transfer, for heat exchanger surfaces, controlling the selection of a given surface. In this chapter we must present experimental data on the pressure drop for finned tubes, obtained as a function of the flow mode, finning geometry and bundle configuration.

5.1. PRELIMINARY REMARKS

Pressure drop for a finned tube bundle is a function of the flow velocity, bundle configuration, finning geometry and physical properties of the coolant. Under the conditions of our experiments, when the coolant flow was virtually isothermal, the hydraulic drag of a bundle, expressed in terms of the pressure drop across it, can be functionally expressed as

$$\Delta p = f(w;\ d;\ s;\ h;\ \delta_1;\ s_1;\ s_2;\ z;\ \mu;\ \rho) \qquad (5.1)$$

Substituting nondimensional ratios into this expression and defining the finning geometry in terms of h/d and s/d, and the bundle configuration in terms of the relative pitches, we obtain:

$$\text{Eu} = \varphi\,(\text{Re};\ h/d;\ s/d;\ s_1/d;\ s_2/d;\ z) \tag{5.2}$$

As shown by a number of previous investigations, the fin thickness has little effect on bundle pressure drop. Also, the fin thickness in the finned tubes under study (see Table 1.2) was constant. Hence Eq. (5.2) does not contain a parameter to account for the fin thickness.

The experimental data for each bundle of a given configuration and finning geometry was correlated by the expression:

$$\text{Eu} = k\,\text{Re}^{-r} \tag{5.3}$$

The flow temperature for each set of experimental conditions was maintained constant. Moreover it varied only between 25 to 75° over the entire range of experiments, for which reason the physical properties with similar criteria were taken at the free-stream temperature.

The contribution of surface friction to the total drag of a finned tube bundle is much greater than in the case of a smooth-tube bundle. However, since the experiments were performed over a relatively narrow temperature range, the additional effect of a boundary layer viscosity variation, associated with different coolant temperatures, on the overall drag of the bundles was insignificant.

Preliminary analysis of the experimental results showed that the bundle pressure drop is proportional to the number of longitudinal rows and is a function of the degree of expansion and of the size of the spaces between the tubes and between the fins in flow direction. The damping out of vortices by the surfaces of the spiral-shaped fins results in a loss of kinetic energy. The first and last rows have the greatest specific kinetic energy loss and this must be taken into account in determining the pressure drop for a small number of rows. In the seven-row bundles under study these additional losses were an insignificant part of the overall bundle drag, and hence the bundle pressure drop was assumed in our calculations to be proportional to the number of rows.

Some bundles with relatively large s_1 comprised only two transverse rows of tubes; the available test section was too small to mount a larger number of transverse rows. When the experiments are properly performed, the results should be independent of the number of transverse rows (i.e., the number of rows normal to the flow direction). This was checked out with a special experiment. We selected a relatively dense bundle, with $(a \times b) = 2.06 \times 1.78$, and with seven longitudinal rows, and we determined its pressure drop with four, three and two transverse rows. The results, reduced in nondimensional form, are plotted in Fig. 5.1, from which it can be seen that the number of transverse rows does not have a perceptible effect on the drag of the bundle—the experimental data for all three bundles lie, within the limits of experimental error, close to one another.

92 HEAT TRANSFER OF FINNED TUBE BUNDLES IN CROSSFLOW

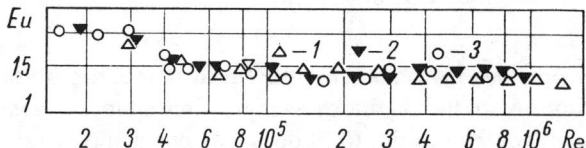

Figure 5.1 Experimentally measured drag of a 2.06 × 1.76 bundle of finned tubes with $d \times s \times h = 23 \times 6.5 \times 10$ with a different number of transverse rows ($z = = 7$ is the number of longitudinal rows). *1)–4)* transverse rows.

5.2. EXPERIMENTAL DATA ON PRESSURE DROP IN TUBE BUNDLES

The experimental results were reduced using Eq. (5.8). The reference dimension here was the diameter of the tube carrying the fins, and the reference velocity was that in the narrowest flow passage, which occurred, depending on the bundle configuration, either along the transverse pitch, or along the diagonal (see Fig. 5.2).

The experimental results were obtained with 21 staggered-bundle configurations. In the first 12 bundles we varied the finning geometry and in the remaining (13 to 21) we changed the arrangement of tubes within the bundle.

Data obtained in this study are presented in the appendices, Tables 2 through 14. The functional curves for these data are given in Figs. 5.3 through 5.23. The constants and exponents of the nondimensional relationships obtained for each bundle are listed in Table 3.

5.3. ANALYSIS OF EXPERIMENTAL DATA

It is seen from the experimental data (Figs. 5.3 through 5.23) that the Euler number for the bundles gradually decreases with increasing Re, until, at some

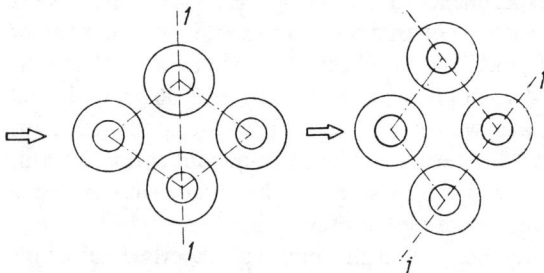

Figure 5.2 Tube arrangements. The least flow passage is along 1-1.

PRESSURE DROP IN FINNED TUBE BUNDLES 93

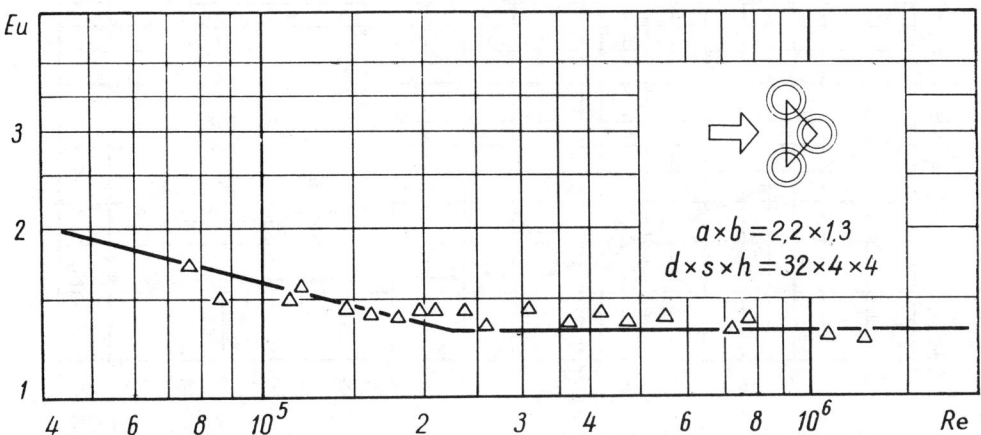

Figure 5.3 Euler number for bundle 1.

value of Re, specific to each bundle and termed by us the "critical Reynolds number," the value of the Euler number becomes constant. This means that there is a fundamental change in the relationship between flow velocity and flow resistance. This suggests that this process can be treated as an analog of the "self-similarity" region, which arises in pipe flows at high Re where there is a nearly constant friction factor and the pressure gradient varies with the square of the velocity.

It was established in our studies of pressure drop in smooth-tube bundles at high Re [20] that transition to self-similar behavior, (i.e., to fully developed external turbulent flow) occurred for all the bundles under study over the Re range from 1×10^5 to 3×10^5. The Reynolds number for transition to self-

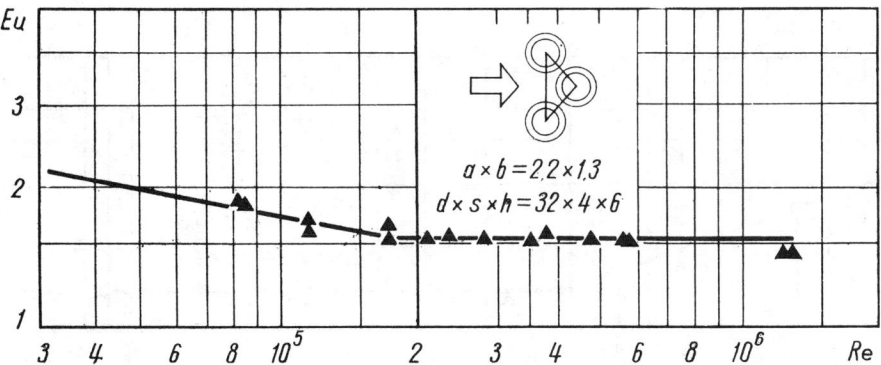

Figure 5.4 Euler number for bundle 2.

94 HEAT TRANSFER OF FINNED TUBE BUNDLES IN CROSSFLOW

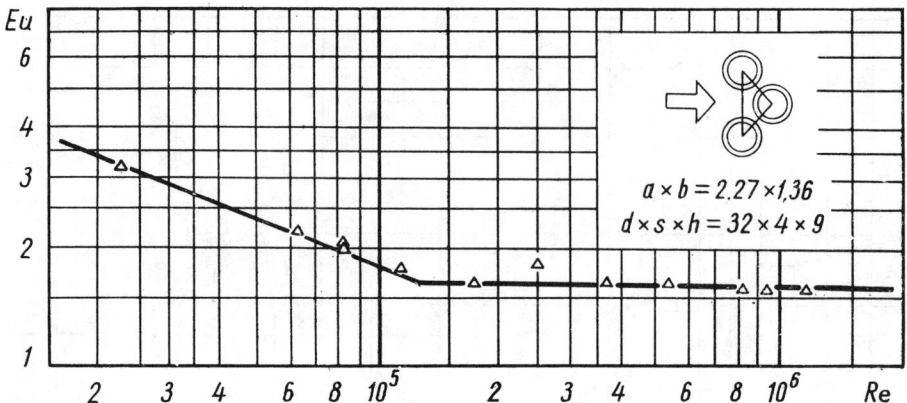

Figure 5.5 Euler number for bundle 3.

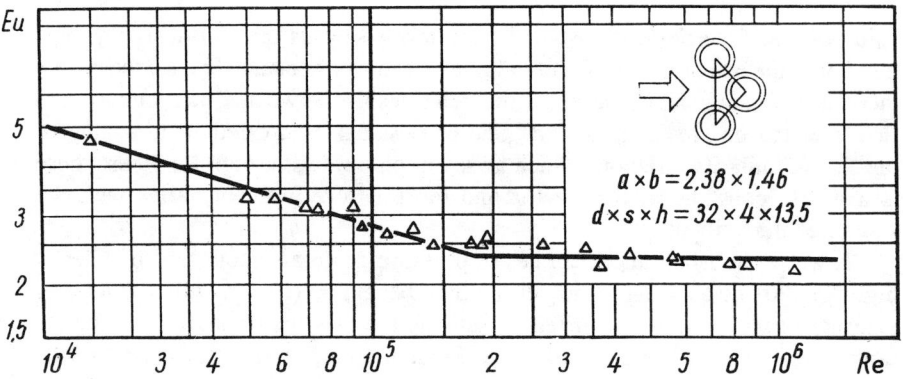

Figure 5.6 Euler number for bundle 4.

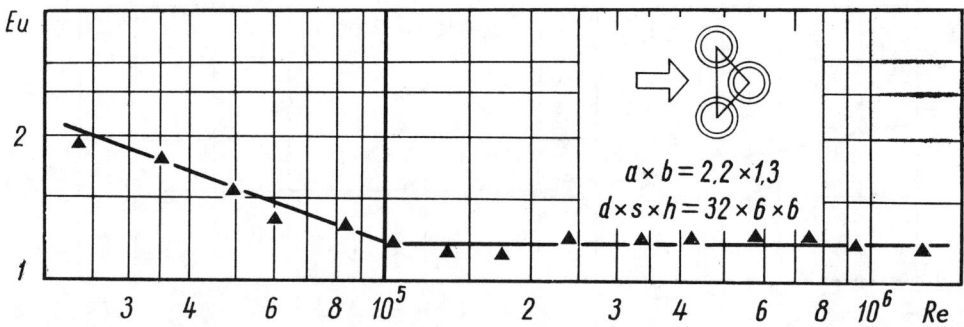

Figure 5.7 Euler number for bundle 5.

PRESSURE DROP IN FINNED TUBE BUNDLES 95

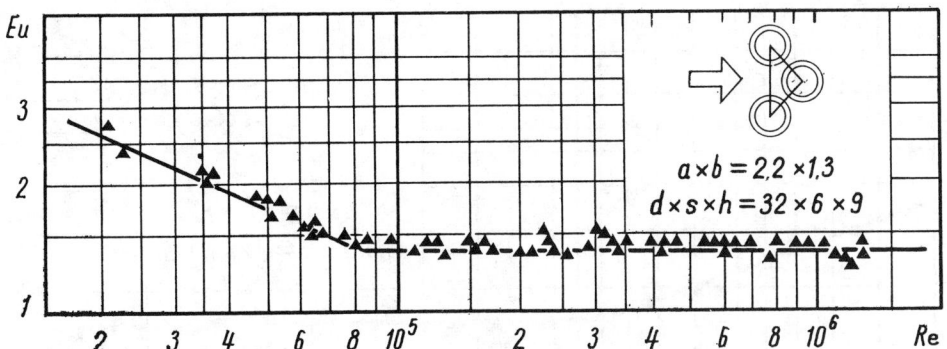

Figure 5.8 Euler number for bundle 6.

similarity depends on the size of the space between tubes. In close-packed bundles ($a/b < 1.7$) self-similarity sets in earlier, whereas in more sparsely packed bundles ($a/b > 1.7$) it occurs at higher values of Re.

The transition to self-similarity in finned tubes is more complicated, due to the existence of a larger number of controlling factors. In our studies, it occurred somewhat earlier than in smooth tube bundles and over an Re range from 6×10^4 to 1.9×10^5. Analysis of experimental results showed that the transition to self-similar behavior occurs at lower Re for at a smaller finned tube diameter ($d = 23$ mm) and with a denser bundle configuration ($a/b < 1.7$, bundles 9, 10, 13, 16 and 17). The smaller tube diameter apparently facilitates a more intense vortex generation. When the diameter of the finned is increased ($d = 32$ mm) and the fin pitch is reduced to 4 mm, transition is delayed to Re ($1.3-1.9 \times 10^5$ even in relatively dense bundles (bundles 1 through 4). Reduc-

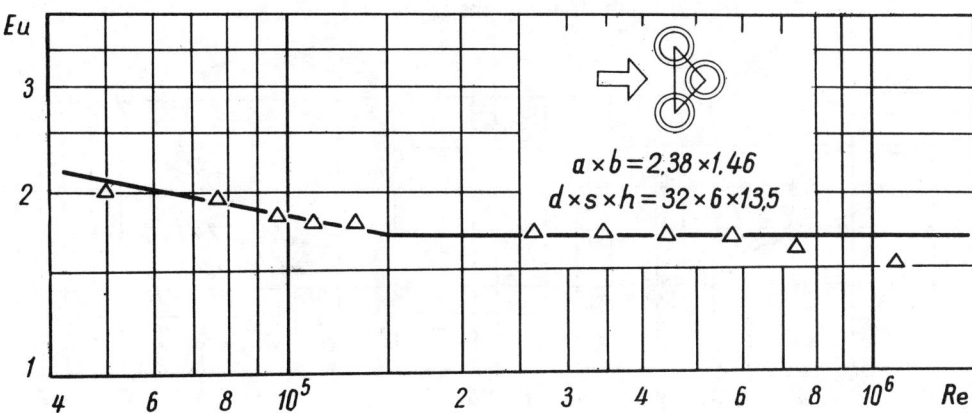

Figure 5.9 Euler number for bundle 7.

96 HEAT TRANSFER OF FINNED TUBE BUNDLES IN CROSSFLOW

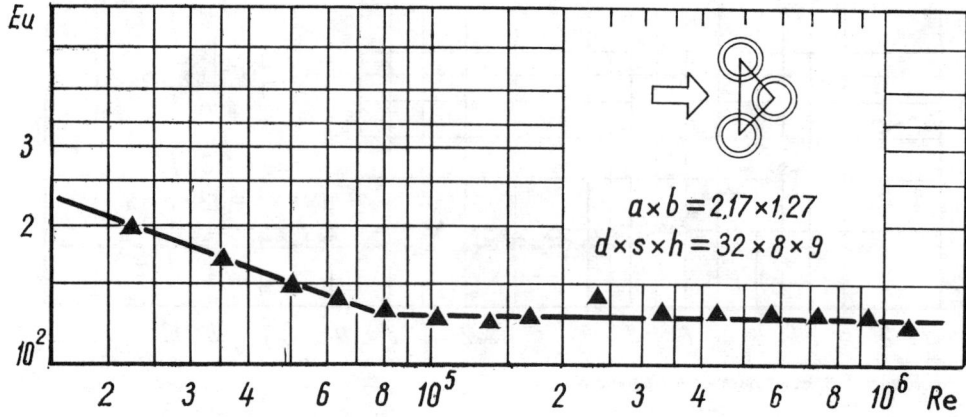

Figure 5.10 Euler number for bundle 8.

ing the fin pitch results in the effective splitting of the flow within the bundle into individual streams which, apparently, facilitates damping of the vorticity and delaying transition to fully developed turbulent flow.

The flow region preceding the onset of fully developed turbulent flow (i.e., below the critical value of Re) is termed, as in smooth-tube bundles, "subcritical." The Euler number for the bundle in this zone is described by a power law. The expressions for Eu = f(Re) for all the bundles under study in both flow regions are listed in Table 3). To illustrate the variation with bundle parameters, the experimental data for bundles 1 through 12 are presented together for each flow region in Figs. 5.24 and 5.25. It can be concluded by examining this information that the Euler number increases with increasing fin height at constant fin spacing at (bundles 1 through 7 and 9 through 11), and increases also with reduction in fin pitch at constant fin height (bundles 8, 6, and 3). This situation is observed in both flow regions (Figs. 5.24 and 5.25).

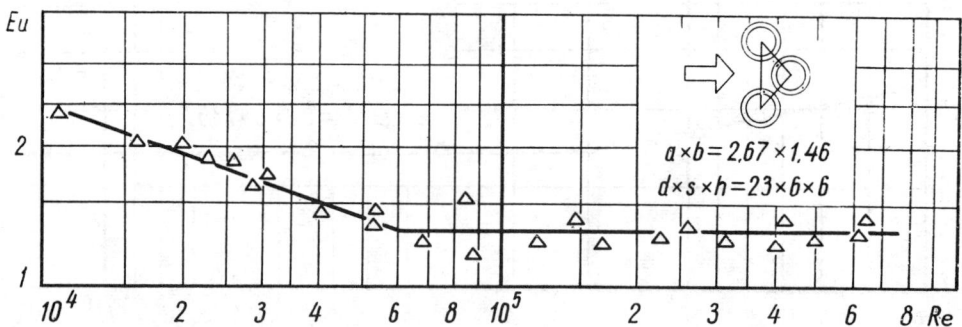

Figure 5.11 Euler number for bundle 9.

PRESSURE DROP IN FINNED TUBE BUNDLES 97

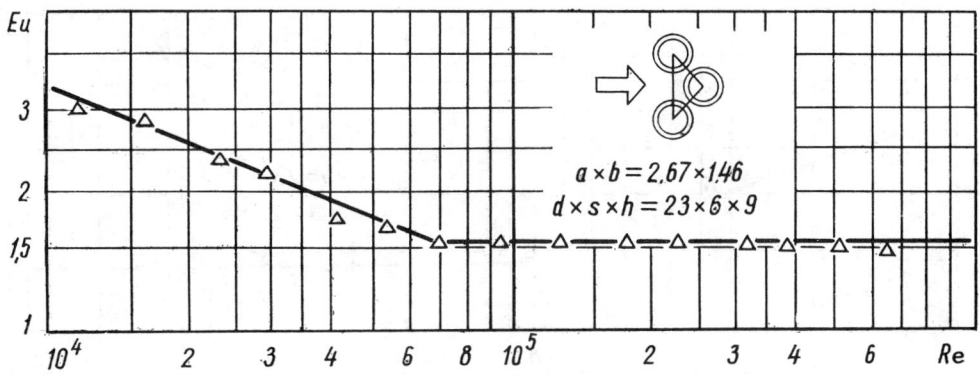

Figure 5.12 Euler number for bundle 10.

It is seen by comparing the data for bundles of different configurations (bundles 13 through 21) under both flow conditions (Figs. 5.26 and 5.27) that the Euler number is directly proportional to the bundle density. The Euler number is highest in bundle 20, which has the highest density ($a \times b = 2.67 \times 1.46$) and, conversely, the lowest is in bundle 18, which is the less densely packed ($a \times b = 4.11 \times 2.14$). The effect of changing the transverse and longitudinal pitches of the bundle on drag is seen by comparing the data for bundles 18, 20 and 21, which differ most significantly by their configuration. It shows that the Euler number for a bundle decreases with an increase in a and b.

It was noted in correlating the previously presented experimental data that the effect of the finning and bundle configuration parameters on the Euler number is approximately the same for both subcritical and supercritical ranges of Re. This is also reflected in the resultant equations. In the correlations of bundle Eu for both flow ranges the effect of fin height is estimated by the

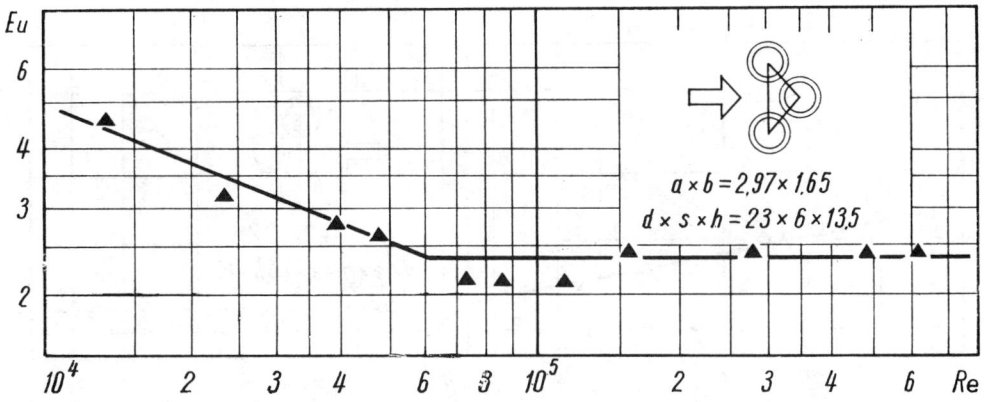

Figure 5.13 Euler number for bundle 11.

98 HEAT TRANSFER OF FINNED TUBE BUNDLES IN CROSSFLOW

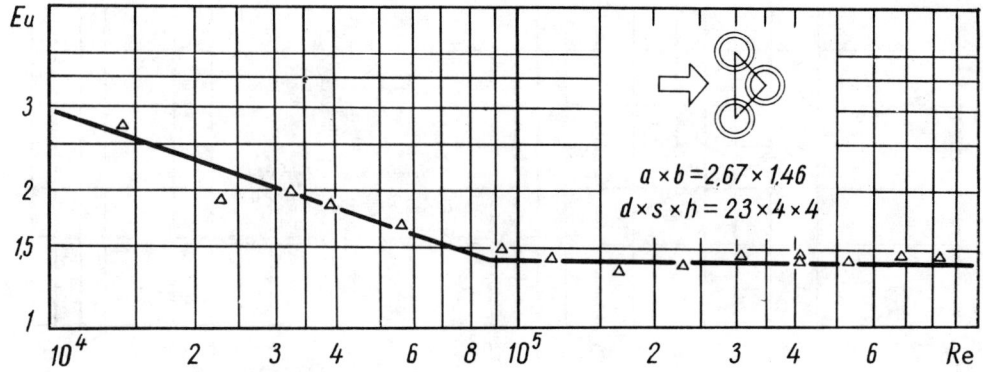

Figure 5.14 Euler number for bundle 12.

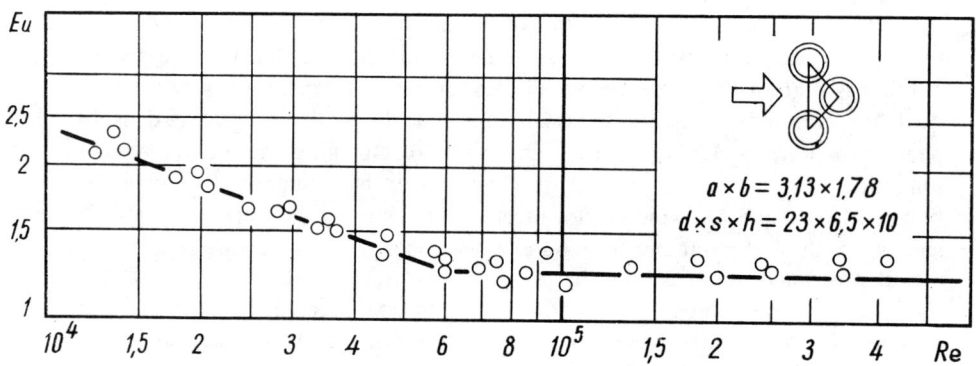

Figure 5.15 Euler number for bundle 13.

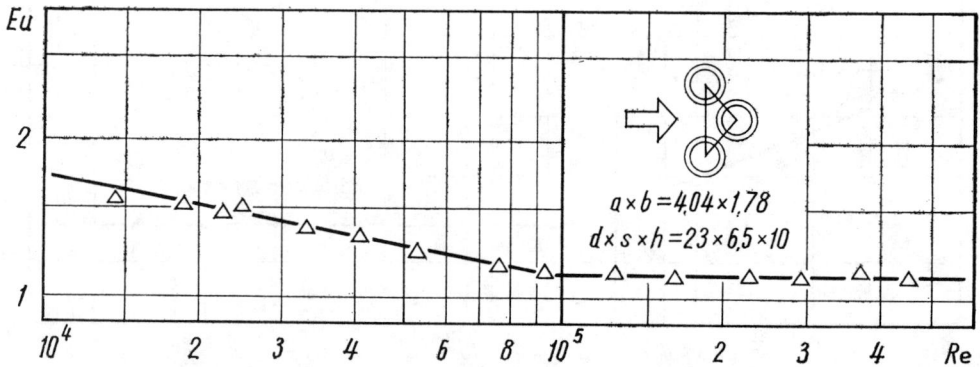

Figure 5.16 Euler number for bundle 14.

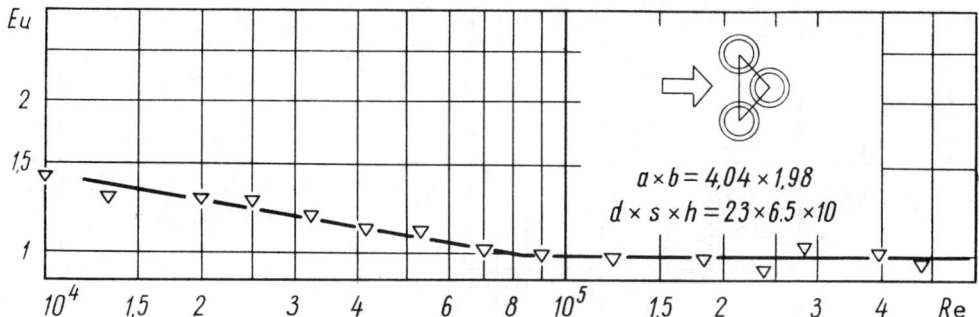

Figure 5.17 Euler number for bundle 15.

parameter $(1 - h/d)^{1.4}$ and that of the fin pitch $(1 - s/d)^{1.8}$. The effect of bundle configuration is reflected respectively by single exponents on the relative pitches ($a^{-0.55}$ and $b^{-0.5}$).

The constant of the correlation equation and the exponent of Re are determined from Figs. 5.28 and 5.29, which are plots of the function

$$K_c = \frac{\text{Eu}\, a^{0.55}\, b^{0.5}\, (1 - h/d)^{1.4}}{z\, (1 - s/d)^{1.8}} = f(\text{Re})$$

for the subcritical and supercritical regions respectively.

It is seen from the graphs that the scatter of points about a single curve does not exceed $\pm 20\%$ for subcritical flow and $\pm 13\%$ for supercritical flow. The critical value of Re assigned for this was found to be Re $\simeq 10^5$.

As a result we obtained the following correlations for determining the Euler number for multirow finned-tube bundles for Re from 10^4 to 10^5

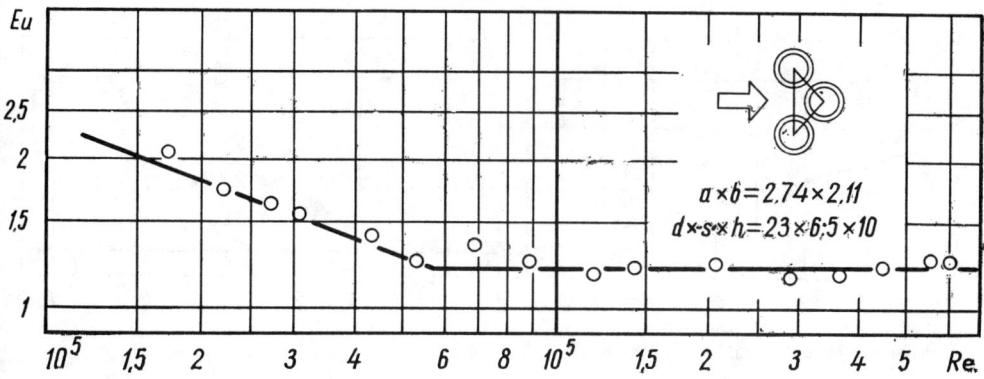

Figure 5.18 Euler number for bundle 16.

100 HEAT TRANSFER OF FINNED TUBE BUNDLES IN CROSSFLOW

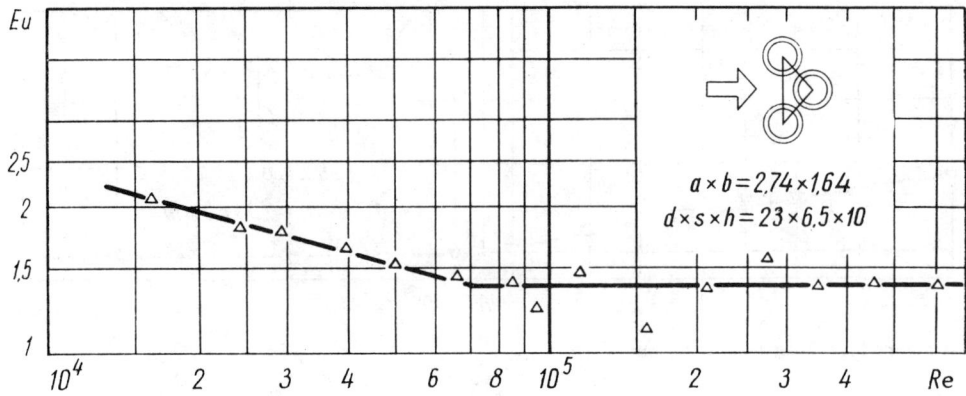

Figure 5.19 Euler number for bundle 17.

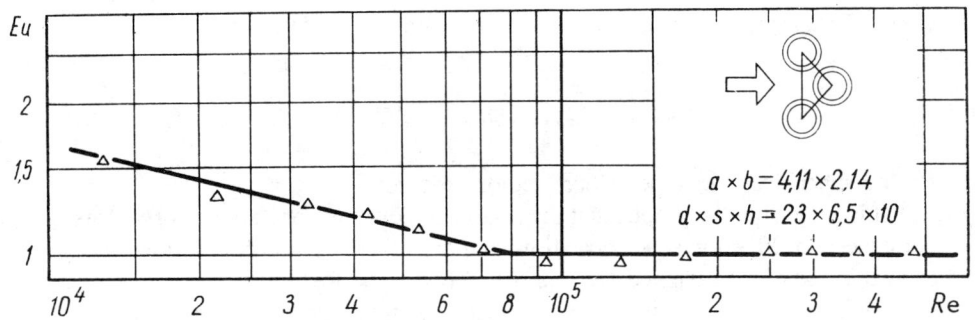

Figure 5.20 Euler number for bundle 18.

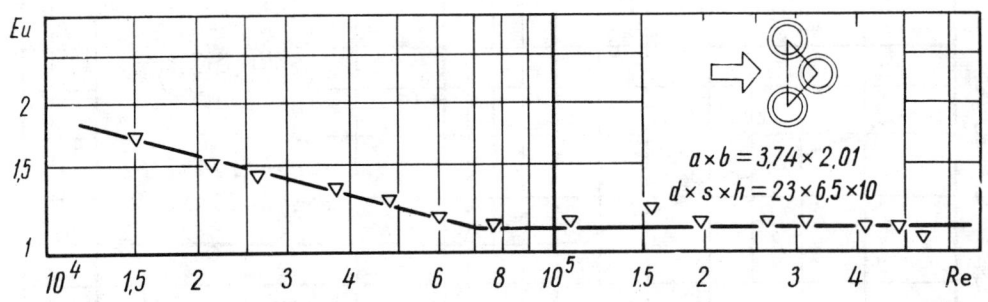

Figure 5.21 Euler number for bundle 19.

PRESSURE DROP IN FINNED TUBE BUNDLES 101

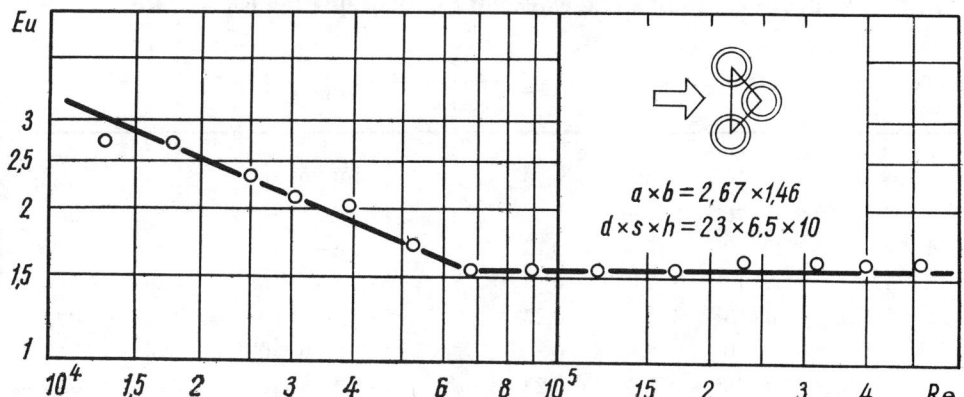

Figure 5.22 Euler number for bundle 20.

$$\mathrm{Eu} = \frac{6.55\,(1-s/d)^{1.8}\,\mathrm{Re}^{-0.25}}{a^{0.55}\,b^{0.5}\,(1-h/d)^{1.4}} \cdot z \qquad (5.4)$$

and for Re from 10^5 to 10^6

$$\mathrm{Eu} = \frac{0.37\,(1-s/d)^{1.8}}{a^{0.55}\,b^{0.5}\,(1-h/d)^{1.4}} \cdot z \qquad (5.5)$$

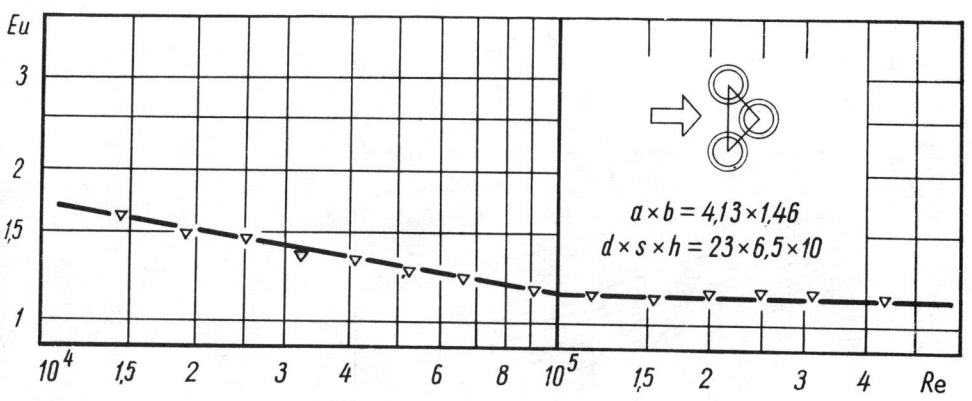

Figure 5.23 Euler number for bundle 21.

TABLE 3 Values of constant k and exponent r in the Equation $Eu = k\,Re^{-r}$

Bundle number	Re range	k	$-r$	Bundle number	Re range	k	$-r$
1	$8.5 \cdot 10^4 - 1.7 \cdot 10^5$	34	0.26	12	$1.4 \cdot 10^4 - 9 \cdot 10^4$	61	0.33
	$1.7 \cdot 10^5 - 1.3 \cdot 10^6$	1.4	0		$9 \cdot 10^4 - 1 \cdot 10^6$	1.4	0
2	$8.5 \cdot 10^4 - 1.9 \cdot 10^5$	15.6	0.19	13	$1 \cdot 10^4 - 6 \cdot 10^4$	60.3	0.35
	$1.9 \cdot 10^5 - 1.3 \cdot 10^6$	1.55	0		$6 \cdot 10^4 - 5 \cdot 10^5$	1.26	0
3	$2.2 \cdot 10^4 - 1.3 \cdot 10^5$	129	0.37	14	$1 \cdot 10^4 - 1 \cdot 10^5$	9.14	0.18
	$1.3 \cdot 10^5 - 1.1 \cdot 10^6$	1.68	0		$1 \cdot 10^5 - 5 \cdot 10^5$	1.13	0
4	$2 \cdot 10^4 - 1.8 \cdot 10^5$	81.3	0.29	15	$1 \cdot 10^4 - 8 \cdot 10^4$	7.61	0.18
	$1.8 \cdot 10^5 - 1.1 \cdot 10^6$	2.35	0		$8 \cdot 10^4 - 5 \cdot 10^5$	1.00	0
5	$2.6 \cdot 10^4 - 9.6 \cdot 10^4$	86.2	0.37	16	$1 \cdot 10^4 - 6 \cdot 10^4$	79.5	0.38
	$9.6 \cdot 10^4 - 8.5 \cdot 10^5$	1.2	0		$6 \cdot 10^4 - 6 \cdot 10^5$	1.22	0
6	$2.2 \cdot 10^4 - 9.5 \cdot 10^4$	155	0.41	17	$1 \cdot 10^4 - 7 \cdot 10^4$	28.2	0.27
	$9.5 \cdot 10^4 - 1.2 \cdot 10^6$	1.4	0		$7 \cdot 10^4 - 6 \cdot 10^5$	1.4	0
7	$5 \cdot 10^4 - 1.5 \cdot 10^5$	14.7	0.18	18	$1 \cdot 10^4 - 8 \cdot 10^4$	33.5	0.26
	$1.5 \cdot 10^5 - 1.1 \cdot 10^6$	1.7	0		$8 \cdot 10^4 - 5 \cdot 10^5$	1.00	0
8	$2.2 \cdot 10^4 - 8 \cdot 10^4$	55.5	0.33	19	$1 \cdot 10^4 - 7 \cdot 10^4$	23.1	0.27
	$8 \cdot 10^4 - 1.2 \cdot 10^6$	1.33	0		$7 \cdot 10^4 - 6 \cdot 10^5$	1.12	0
9	$1 \cdot 10^4 - 6 \cdot 10^4$	59.5	0.35	20	$1 \cdot 10^4 - 7 \cdot 10^4$	94.5	0.36
	$6 \cdot 10^4 - 6 \cdot 10^5$	1.3	0		$7 \cdot 10^4 - 5 \cdot 10^5$	1.65	0
10	$1 \cdot 10^4 - 7 \cdot 10^4$	146	0.41	21	$1 \cdot 10^4 - 1 \cdot 10^5$	8.94	0.18
	$7 \cdot 10^4 - 6.5 \cdot 10^5$	1.55	0		$1 \cdot 10^5 - 5 \cdot 10^5$	1.16	0
11	$1.3 \cdot 10^4 - 6 \cdot 10^4$	195	0.4				
	$6 \cdot 10^4 - 8 \cdot 10^5$	2.4	0				

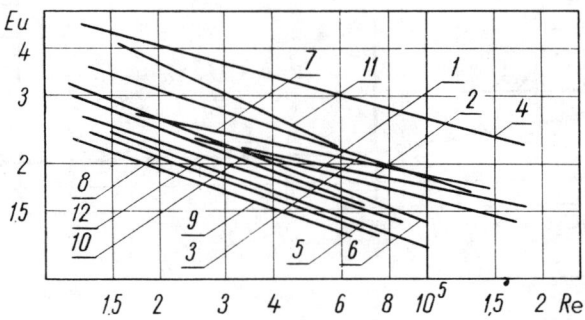

Figure 5.24 Euler number for bundles 1 through 12 in the mixed flow region.

PRESSURE DROP IN FINNED TUBE BUNDLES **103**

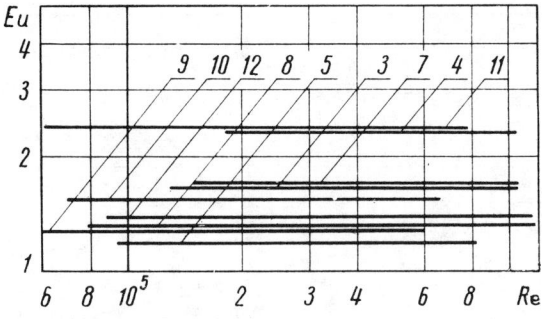

Figure 5.25 Euler number in the developed turbulent flow region.

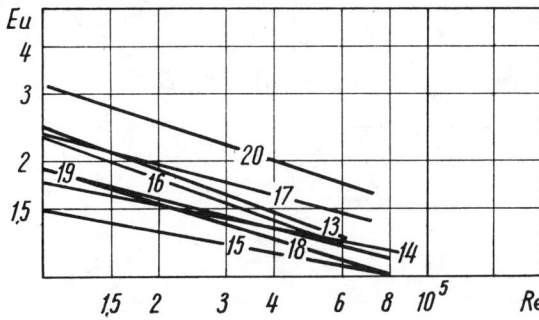

Figure 5.26 Euler number in the mixed-flow region.

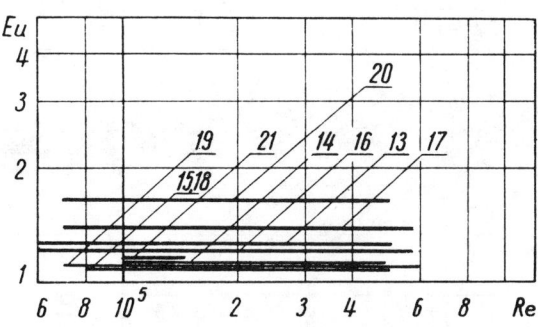

Figure 5.27 Euler number curves for bundles 13 through 21 in the fully developed turbulent flow (supercritical) region.

104 HEAT TRANSFER OF FINNED TUBE BUNDLES IN CROSSFLOW

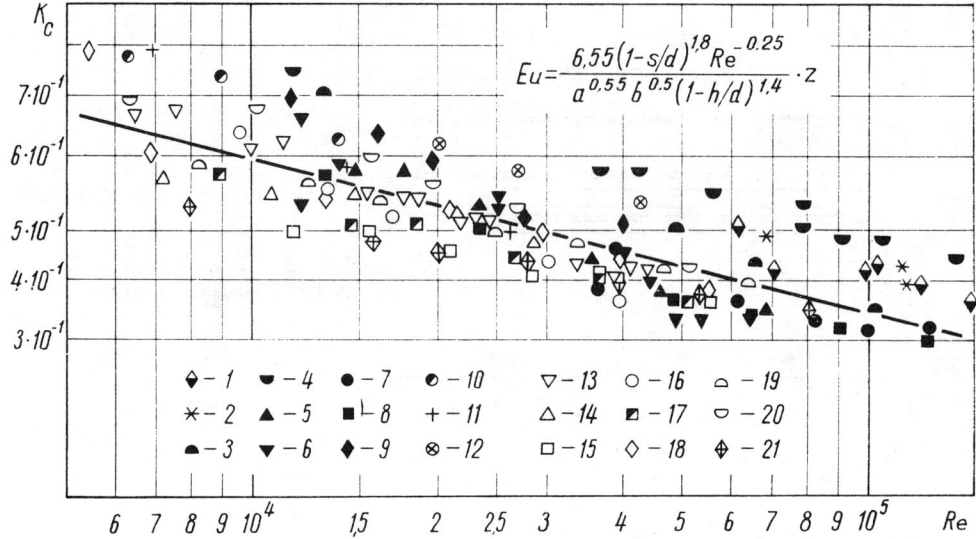

Figure 5.28 Plot of $K_c = f(\mathrm{Re})$ for the subcritical flow region.

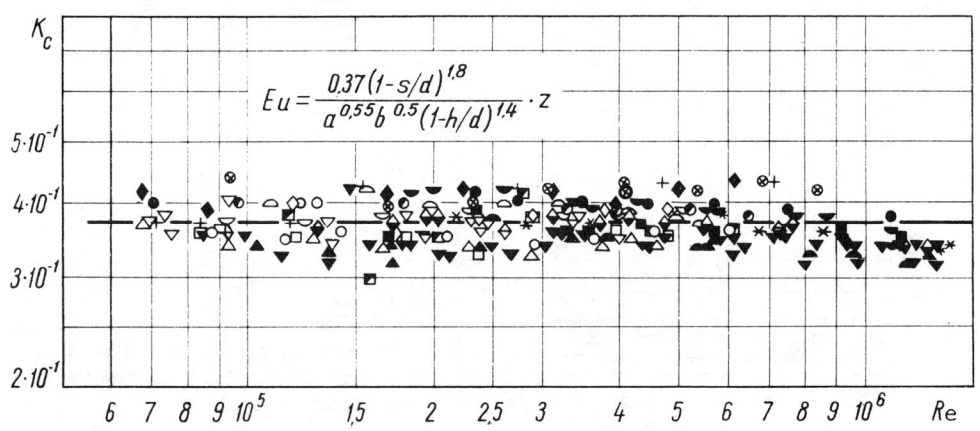

Figure 5.29 Plot of $K_c = f(\mathrm{Re})$ for the supercritical flow region.

CHAPTER
SIX

EXPERIMENTAL DETERMINATION OF THE DISTRIBUTION OF THE HEAT TRANSFER COEFFICIENT OVER A FIN

It is known that working equations obtained representing experimental data for a specific tube, and obtained on the basis of correlating the reduced heat transfer coefficient for the given tube, have the shortcoming of not being suitable for use in correlating data for the same geometry but with a tube material of different thermal conductivity. On the other hand, experimental data based on convective heat transfer coefficients lend themselves easily to correlation, but can be determined experimentally only with difficulty. This dilemma was resolved by working out a method for determining the convective heat transfer coefficient on the basis of the reduced heat transfer coefficient, the fin effectiveness and a coefficient describing the nonuniformity of heat transfer over the fin surface, contained in Eq. (3.7).

The present chapter is concerned with a detailed analysis of this question. It describes special experiments for determining the distribution of heat transfer over the surface of a straight circumferential fin and a trapezoidal fin.

6.1. STATEMENT OF THE PROBLEM

The analytic expressions for the fin effectiveness presented in Chapter 2 were obtained by making a number of assumptions. Some of these assumptions are unimportant in their effect on the results obtained; for instance, the likely deviations from the assumptions of a constant fin base temperature and a negligible temperature gradient perpendicular to the fin surface, are usually such as to have little effect on the final answer. On the other hand, some of the other assumptions must be examined critically; these include neglect of the amount of heat transmitted through the tip surface of the fin and the assumption of a constant heat transfer coefficient over the fin. The heat transmitted at the tip cannot be disregarded, since it comprises a significant fraction of the entire heat transmission (in the finned tubes under study, it amounted to up to 15%). However, as mentioned above, heat transfer from the tip of a fin can be accounted for rather precisely by conditionally increasing fin height by one half the fin width. The assumption of a constant heat transfer coefficient over the fin surface is also unjustified, as was shown in the preceding chapter. According to actual measurements, the heat transfer coefficient increases in the direction of the fin tip. In straight plane fins the ratio of heat transfer coefficient between tip and base is approximately 1.3 to 1.4; in circular fins it is somewhat smaller [5, 55]. Nonuniformity in distribution of the heat release rate over a finned surface may be demonstrated by measuring the distribution of the local heat heat transfer coefficient over the fin surface. According to experimental data obtained by us and by others, and presented in Sec. 8.2, local heat transfer coefficients are nonuniform over the fin surface and are higher at the fin tip than at the base. This nonuniformity reduces the fin effectiveness to below that calculated using a mean heat transfer coefficient α, assumed to be constant over the entire finned surface. Thus, in analyzing data for finned surfaces, it is necessary to make a correction to take account of the heat transfer coefficient nonuniformity on fin effectiveness.

The first attempt in this direction was made by Karasina [13]. She introduced the correction factor ψ, incorporating the nonuniformity of distribution of the heat transfer coefficient, ψ being assumed to be constant over the range of βh from 0.1 to 1.1, and equal to 0.85. As a result, she suggested that Eq. (3.9) be written in the form:

$$\alpha_{red} = \left(\frac{F_{fin}}{F} E \xi + \frac{F_t}{F} \right) \alpha \psi \qquad (6.1)$$

where ξ is a coefficient taking account of the variation in fin thickness over its height (see Fig. 3.3). In this case the value of E referred to the function

$$h \sqrt{\frac{2\alpha\psi}{\delta\lambda}}$$

Doubts raised concerning the constancy of the expression $\psi = f(\beta h)$ were confirmed in subsequent studies.

Migay [45] specified arbitrarily different distributions of the heat transfer coefficient over the height of a plane straight fin, and determined analytically the magnitude of the correction to the expression

$$\alpha_{red} = \alpha E \Pi \tag{6.2}$$

where π is a correction factor for nonuniformity of the heat transfer coefficient, and is defined as

$$\Pi = \frac{Q_{\tilde{\alpha}}}{Q_{\bar{\alpha}}} \tag{6.3}$$

where $Q_{\tilde{\alpha}}$ is the amount of heat transmitted through a plane fin and determined on the basis of a variable heat transfer coefficient, and $Q_{\bar{\alpha}}$ is the amount of heat transmitted through the same fin at the mean heat transfer coefficient. The nonuniformity of the heat transfer coefficient was defined as $\epsilon = \alpha_b/\alpha_0$, where α_b and α_0 are the heat transfer coefficients at the fin tip and fin base respectively.

Figure 6.1 shows diagrams of the assumed distributions of the local heat transfer coefficient over the fin. Versions I and II correspond to linear variation in α over the fin height, III and IV to a hyperbolic variation in this coefficient. The correction coefficients thus obtained (see Fig. 6.2) show that they vary differently as a function of the nondimensional fin height βh. The principal effect on the variation in Π is exerted by the nature of the distribution over the fin height. Depending on the latter, Π can be both smaller and greater than unity. It follows from the work by Migay [45] that at $\beta h \to 0$ for any nonuniformity of heat transfer over the fin height we have $\Pi \to 1$. Over the range $1 >$

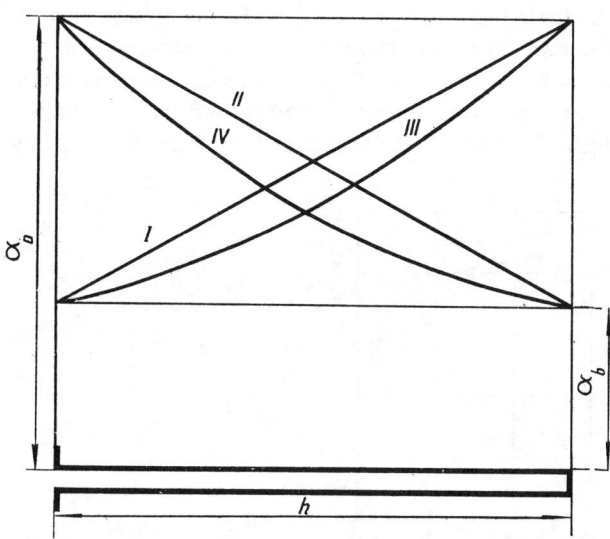

Figure 6.1 Arbitrary distributions of heat transfer coefficients over the height of a fin assumed in the analysis of Migay (45).

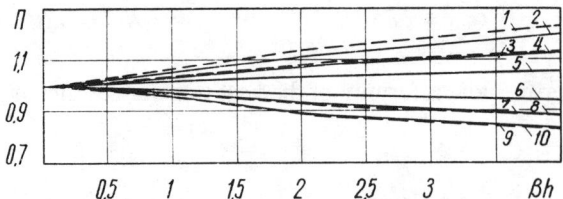

Figure 6.2 Effect of nonuniformity of heat transfer over the height of a plane fin Π for different ratios α_0/α_b and for the different forms of distribution showin in Fig. 6.1. *1)* α_0/α_b = 3.3, version IV; *2)* 3.3 and II; *3)* 2 and IV; *4)* 2 and II; *5)* 1.25 and IV; *6)* 0.8 and I and III; *7)* 0.5 and I; *8)* 0.5 and III; *9)* 0.3 and I; *10)* 0.3 and III.

$\beta h > 0$ the effect of nonuniformity is insignificant (the correction does not exceed 6%). This is due to the fact that at these values of βh the thermal resistance of the fin material is insignificant compared to the thermal resistance of the thermal boundary layer on the fin surface. The fin thermal resistance becomes more important with increasing βh, which is the reason for the increase in Π with increasing βh. According to Migay [45], the nonuniformity effect should be greater in the case of cylindrical transverse fins, as follows from Fig. 6.2, since consideration must additionally be given to the nonuniformity of the fin area over its height.

Brauer [46] also suggested that a correction factor be used in estimating the effectiveness of a plane rectangular fin. He noted in reducing his experimental data that, as a rule, the value of α_0 is lower at the base than at the tip of the fin. The most probably ratio of the minimum to the maximum heat transfer coefficients is, according to Brauer, $0.3 > \alpha_0/\alpha_b < 0.6$. He suggested that the correction factor to be the ratio of fin effectiveness values determined at variable and constant convective heat transfer coefficients α on the fin surface over its height $\Pi = E_{\tilde{\alpha}}/E_{\bar{\alpha}}$. It follows, for example, from Fig. 6.3 that for $\beta h = 1$ (which has corresponding to it $E\alpha = 0.76$) and over the range of α_0/α_b cov-

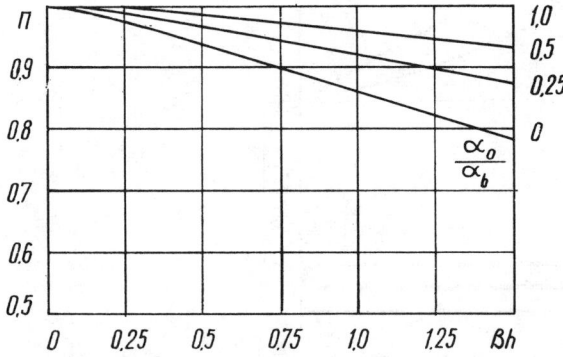

Figure 6.3 Correction factor $\Pi = E_{\tilde{\alpha}}/E_{\bar{\alpha}}$ vs βh and α_0/α_b.

ered, Π varies from 0.93 to 0.97. At higher fin effectiveness, Π naturally, will approach 1.

Sasin [47] developed further this idea and also noted, by analyzing the results of various investigators that the ratio of the minimum to the maximum heat transfer coefficient over a fin ranges, on average, between 0.3 and 0.7, with minimum values at the fin base. He suggested a graphical method for determining Π for a right circumferential fin over a wide range of α_0/α_b. For a given α_0/α_b, Π was given as a function of the product

$$\beta h \sqrt{\frac{D}{d}}$$

(Fig. 6.4). For $\alpha_0/\alpha_b = 0$, 0.25 and 0.5 the graph was constructed from data of Brauer [46], and for other values of this ratio, from the complex equation suggested by Sasin [47]

$$\Pi = 1.05 - \left[0.085 \left(1.42 - \frac{\alpha_0}{\alpha_b}\right)^2 - 0.015\right] \beta h \sqrt{\frac{D}{d}} \qquad (6.4)$$

where for $Re < 7 \times 10^4$

$$\frac{\alpha_0}{\alpha_b} = 1 - 0.35 \left(\frac{h}{u}\right)^{0.333}$$

and for $Re\ (7\text{--}9) \times 10^4$

$$\frac{\alpha_0}{\alpha_b} = \left[1 - 0.35 \left(\frac{h}{u}\right)^{0.333}\right] + \left[0.35 \left(\frac{h}{u}\right)^{0.333} - 0.23 \left(\frac{h}{u}\right)^{0.5}\right] \frac{Re - 7 \cdot 10^4}{2 \cdot 10^4}$$

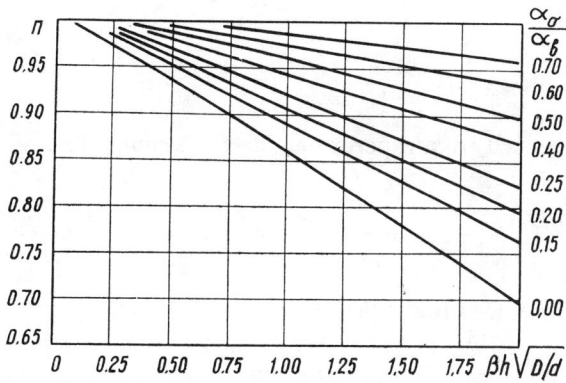

Figure 6.4 Correction factor Π vs $\beta h \sqrt{D/d}$ at different α_0/α_b.

and for $Re > 9 \times 10^4$

$$\frac{\alpha_0}{\alpha_b} = 1 - 0.23 \left(\frac{h}{u}\right)^{0,5}$$

It is seen from Eq. (6.4) that Π depends on Re in addition to

$$\beta h \sqrt{\frac{D}{d}}$$

Thus, Sasin assumes that

$$\Pi = f(\beta h \sqrt{D/d}, \text{Re})$$

All the above pertains to cases when heat transfer from a finned tube occurs at a given values of thermal conductivity for the fin material and the coolant. An alternative approach is to consider the variation of fin effectiveness with fin-metal and coolant thermal conductivity.

Fortescue and Hall [26] concluded, on the basis of their experimental data, that in order to take account of fin-metal and coolant thermal conductivity, for a given finning geometry, one should introduce the ratio λ_f/λ_w. this means that now we have the functional relationship:

$$E = f(\text{Re}^m, \text{Pr}^n, \lambda_f/\lambda_w) \tag{6.5}$$

Lymer and Ridal [51], starting with the extended form of Eq. (2.72) for a longitudinal fin (developed by Dusinberre [74]) and using the data of Fortescue and Hall [26], suggested an approximate solution of this equation, suitable for determining the effectiveness of a right circumferential fin:

$$E = \frac{1}{1 + M \frac{2h^2 \alpha}{3\delta \lambda_w}} \tag{6.6}$$

where M is a coefficient determined from experimental data. Assuming that

$$\alpha E = \frac{\alpha}{1 + M \frac{2h^2 \alpha}{3\delta \lambda_w}} \tag{6.7}$$

and regrouping, Lymer and Ridal [51] found that:

$$\frac{\lambda_f}{\alpha E d} = \frac{\lambda_f}{\alpha d} + M \frac{\lambda_f}{\lambda_w} \frac{2h^2}{3\delta d} \tag{6.8}$$

If the experimental data, obtained at different λ_w, are reduced and plotted in the form

$$\lg \frac{1}{\text{Nu}_{\text{red}}} = f\left(\lg \frac{\lambda_f}{\lambda_w} \frac{2h^2}{3\delta d}\right)$$

then the slope of the resultant curve yields the value of M. Extrapolating this straight line until it intersects with the ordinate axis, i.e., at

$$\frac{\lambda_f}{\lambda_w} \frac{2h^2}{3\delta d} = 0$$

(which would occur at $\lambda_w = \infty$), we obtain the value of $1/\text{Nu}$, from which one can easily calculate the convective coefficient of a finned tube with a given geometry. Lymer and Ridal concluded that at $\lambda_w = \infty$, $\text{Nu}_{\text{red}} = \text{Nu}$ and $E = 1$.

This means that for a given finned-tube geometry and Re, one can obtain heat transfer data for a range of values of both λ_w and λ_f and then use Eq. (6.8) to reduce this data to obtain the heat transfer coefficient for infinite fin thermal conductivity. The values for any given combination of thermal conductivities can then be calculated, again via Eq. (6.8).

Lymer and Ridal determined experimentally the heat transfer coefficient (α_{red}) from tubes with circumferential fins, fabricated from copper, aluminum, carbon and stainless steel. Tubes with $d \times s \times h = 19 \times 12.7$ were placed in a staggered bundle (1.8×1.36). Figure 6.5 shows the results of this study, reduced in the previously described manner at different values of Re. It is seen

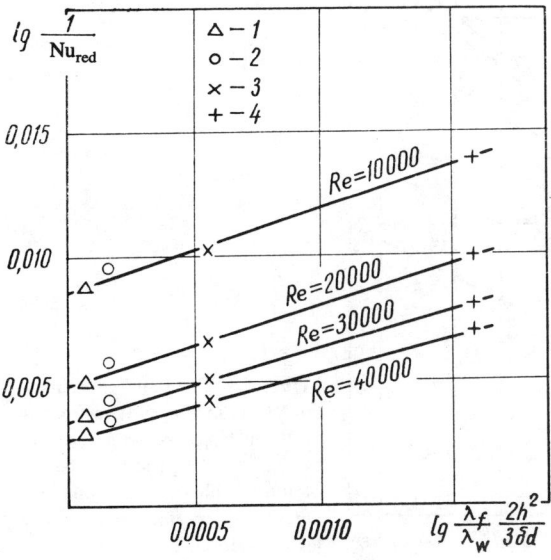

Figure 6.5 Experimental data on heat transfer from bundles of finned tubes, machined from different materials, reduced in the form

$g(1/\text{Nu}_b) = \left[\lg\left(\frac{\lambda_f}{\lambda_w} \frac{2h^2}{3\delta d}\right)\right]$

1) Copper; 2) aluminum; 3) carbon steel; 4) stainless steel.

that the slope *(M)* of the resultant straight lines changes somewhat as a function of Re, thus confirming the dependence of E on Re. Extrapolation of these straight lines to the ordinate axis (at $\lambda_w = \infty$) yields the value of the heat transfer coefficient for infinite fin thermal conductivity, $\alpha_{red} = \alpha$. The data were also plotted in the form $Nu_{red} = f(RE)$; the resultant plot is shown in Fig. 6.6. As will be seen, Nu_{red} depends strongly on the material used, being higher the higher the thermal conductivity. The upper line in Fig. 6.6 represents the extrapolated relationship for infinite conductivity.

Figure 6.7 compares the fin effectiveness obtained by the experimental/graphical method described above and that obtained analytically using the Gardner (uniform α) method. Whereas at high λ_w (with copper fins) the values of effectiveness obtained by the two methods are in satisfactory agreement, than at lower values, and specifically for metals with low thermal conductivity, there exists a significant difference between the two. In fins made of metal with high thermal conductivity the temperature is more uniformly distributed over the height of the fin (i.e., α_{red} approaches α) and the effect of any nonuniformity of the heat transfer coefficient on fin effectiveness is not large. At lower λ_w the temperature distribution over the fin surface is highly nonuniform, and the effect of nonuniformity of α is significant and a special correction is required. This is in satisfactory agreement with conclusions drawn in earlier works.

The ideas of Lymer and Ridal were developed further by Pshenisov and Lukhnov [52], who studied the effect of nonuniformity of the heat transfer coefficient with tubes with spiral fins ($d \times s \times h = 16 \times 4 \times 6$), constructed of

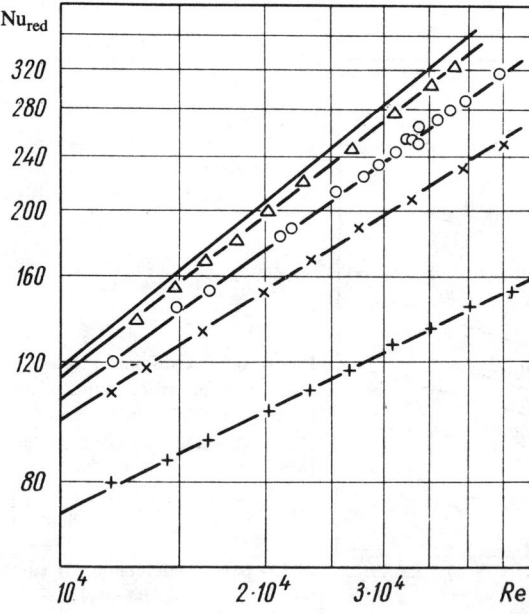

Figure 6.6 The same experimental data as in Fig. 6.5, reduced in the form $Nu_{red} = = f(RE)$. The top curve corresponds to a bundle of tubes with the same fins, but with $\lambda_w \rightarrow \infty$ (for the remaining legend see Fig. 6.5).

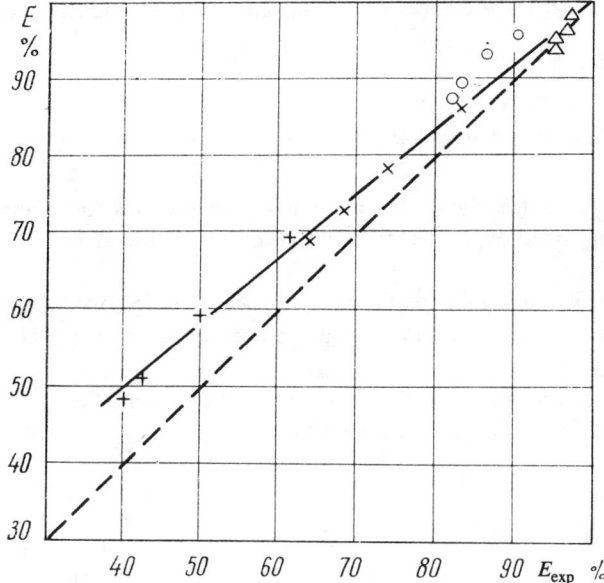

Figure 6.7 Comparison of fin effectiveness values obtained analytically (dashed line) and experimentally (for legend see Fig. 6.5).

different materials and assembled into a staggered bundle ($a \times b = 1.9 \times 2.1$).

Pshenisov and Lukhnov [52] constructed curves of $\mathrm{Nu}_{red} = f(1/\lambda_w)$, extrapolated to $1/\lambda_w = 0$, and obtained the values of the convective heat transfer coefficient. On the basis of these data they determined the values of nonuniformity factor ψ, which are plotted in Fig. 6.8 as a function of βh and Re over the range of λ_w under study.

It is seen from this figure that the reduction in ψ with increasing βh is also a function of Re, and the higher the latter, the smaller ψ, all other conditions being equal. In addition, Pshenisov and Lukhnov suggest that ψ depends, not

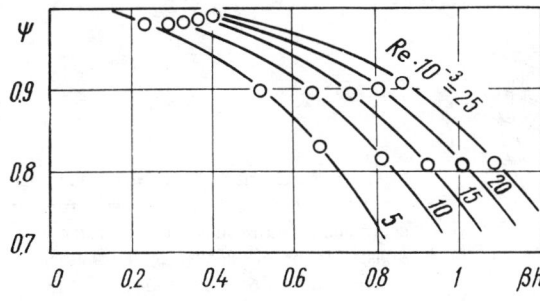

Figure 6.8 Heat transfer nonuniformity factor ψ vs βh at different Re.

only on βh and Re, but also on the ratio D/d. This led them to suggest the expression

6.2. METHODOLOGICAL REMARKS

It follows from the data presented above that no reliable correlations for ψ are available, and that the suggested experimental method (extrapolation G $1/\lambda_w. = 0$) is highly time consuming.

In our study the coefficient ψ of nonuniformity of heat transfer over a fin was determined by a much simpler experimental method using a completely heated model.

For this we constructed two calorimeter tubes: one with spiral and the other with circumferential fins. The tube geometries were:

diameter of tube carrying the fins	$d = 32$ mm,
fin height	$h = 9$ mm,
fin pitch	$s = 6$ mm,
fin thickness at the base	$\delta_1 = 2$ mm,
fin thickness at the tip	$\delta_2 = 1$ mm,
helix angle of spiral fin	$2°2'-4°$

The calorimeter tubes were constructed in such a manner that it would be possible to measure the convective heat transfer coefficient. The design of such a tube is shown schematically in Fig. 6.9. A thin-walled, mica insulated stainless-steel sleeve, which served as a heater, was placed in a copper sleeve which, in turn, was press fitted into a model of a finned tube, machined from steel. The inner space was filled with a mixture of fiberbrick clay and liquid glass. The voltage drop was measured by welding two thoroughly insulated stainless steel strips in the middle part of the heater at 100 mm from one another. The entire tube was

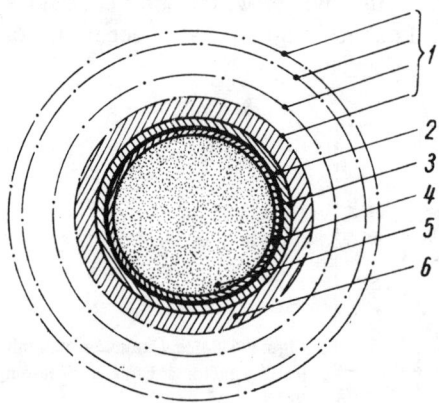

Figure 6.9 Schematic of experimental tube for determining the convective heat transfer coefficient. *1)* Resistance thermometer locations; *2)* copper sleeve; *3)* insulation; *4)* heater; *5)* filler; *6)* body of tube.

HEAT TRANSFER COEFFICIENT OVER A FIN 115

heated by direct current. The mean temperature of the finned tube surface was measured by placing resistance thermometers in five locations around the circumference and at radial positions corresponding to the tube surface, two positions on the fin and the fin tip respectively (see Fig. 6.9). A weighted mean temperature t_m of the finned tube was calculated from the surface thermocouple readings; this mean was determined by multiplying each reading by the surface area to which it referred and dividing by the total surface area. The calorimeter tubes were placed in longitudinal rows 1 and 5 of two staggered-tube bundles, assembled respectively of tubes with spiral or circumferential fins. The bundle configuration in both cases was $a \times b = 2.2 \times 1.3$.

The average convective heat transfer coefficients (α) from the finned tube were obtained from the expression:

$$\alpha = \frac{Q}{F(t_m - t_f)} \qquad (6.9)$$

The equivalent reduced heat transfer coefficients were obtained from Eq. (4.1), with temperature t_1 at the fin base serving as the reference temperature of the surface of the finned tube. Here, t, was calculated as an average of the values of the five circumferential locations.

6.3. EXPERIMENTAL RESULTS

Our experimental data (see appendices, Table 32), are plotted in the form $\alpha_{red} = f(\alpha)$, in Fig. 6.10. The solid curve in this was calculated from the analytical expression

$$\alpha_{red} = \alpha \left[\frac{F_{fin}}{F} E\xi + \frac{F_{t,i}}{F} \right] \qquad (6.10)$$

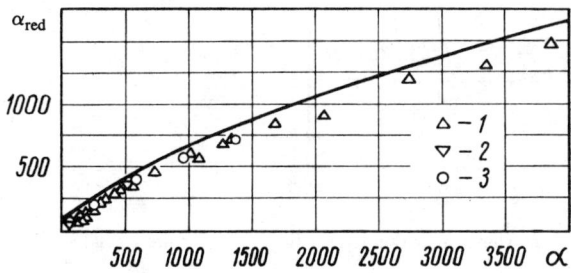

Figure 6.10 Comparison of measured convective heat transfer coefficients with their counterparts calculated from Eq. (6.10). *1)* Spiral fin in fifth row; *2)* spiral fin in first row; *3)* circumferential fin in fifth row. The solid curve is plotted from Eq. (6.10).

from which the convective heat transfer coefficients α were calculated from arbitrarily specified values of α_{red}.

It is seen from this graph that the experimental points do not fit the analytical curve. This is primarily because of the effect on fin efficiency E of the nonuniformity of the distribution of the heat transfer coefficient over the fin. This effect can be incorporated in Eq. (6.10) by including a factor ψ in the first bracketed term (referring to the fin surface). Eq. (6.2) then becomes:

$$\alpha_{red} = \alpha_1 \left[\frac{F_{fin}}{F} E \psi \xi + \frac{F_t}{F} \right] \qquad (6.11)$$

If we calculate values of ψ from this equation for each experimental point and plot them as a function of βh, we obtain an equation in the form $\psi = f(\beta h)$ (Fig. 6.11). It is seen from the figure that over the range $1 > \beta h > 0.3$, the effect of the heat transfer coefficient nonuniformity is insignificant (the correction does not exceed 9%). According to Migay [45] this is because, at these values of βh, the thermal resistance of the fin material is lower than the thermal resistance of the thermal boundary layer on the fin surface. The fin thermal resistance becomes increasingly significant with increasing βh, which causes the ψ to decrease. The circumferential nonuniformity of the heat transfer coefficient was taken into account in the averaging of the readings of the resistance thermometers, as described above.

Fitting the points in Fig. 6.11 with a straight line, we obtain the following expression for ψ:

$$\psi = 0.97 - 0.056 \beta h \qquad (6.12)$$

In the range of βh from 0.3 to 3, the value of ψ decreases from 0.95 to 0.8. It can be assumed that, if βh tends to 0, ψ should tend to 1. However, the straight

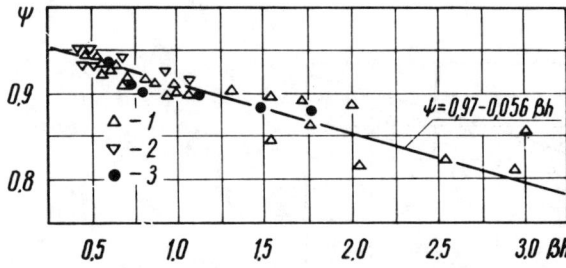

Figure 6.11 Nonuniformity of heat transfer distribution over a fin vs βh (for the legend see Fig. 6.10).

line in Fig. 6.11, obtained by us for $\beta h > 0.3$ extrapolates to 3% lower than unity at $\beta h = 0$.

Equation (6.12) for the correction factor ψ can be used for the more precise determination of the convective heat transfer coefficient of various finned tubes in crossflow.

CHAPTER
SEVEN

MEAN HEAT TRANSFER COEFFICIENT FROM A FINNED TUBE IN A BUNDLE

Equation (3.2) incorporates the principal factors controlling heat transfer from a finned tube in a bundle; these are the similitude criteria, controlling the flow mode and properties of the free stream, and also nondimensional ratios, reflecting the geometries of the finning and of the bundle configuration. The present chapter describes experimental results demonstrating the effect of these factors in the Reynolds number range from 10^4 to 10^6.

Preliminary correlation of experimental data for a single bundle was performed with a power-law equation, based on the measured reduced heat transfer coefficient:

$$\text{Nu}_{\text{red}} = c\text{Re}^m \tag{7.1}$$

Below we present a detailed description of the reduction of experimental data for 21 staggered bundles. Twelve of these data sets were used to investigate the effect on heat transfer of the finning parameters, and the remaining data sets were used to elucidate the effect of tube arrangement.

7.1. RESULTS OF EXPERIMENTAL STUDY OF BUNDLES OF TUBES WITH VARIOUS FIN GEOMETRIES

The experimental results pertain to the study of heat transfer from 12 staggered bundles with different finning geometry. Detailed data on these are given in

Table 1. It is seen from the table that the bundles are arranged in groups. The first group is comprised of bundles with $s = 4$ mm and variable fin height h. The second group is made up of bundles with $s = 6$ mm, and the third with $s = 8$ mm. The fourth and fifth groups are formed of tubes with $d = 23$ mm.

The bulk of the experimental data is presented in tables 15 through 19 in the Appendix. The data, reduced as $Nu_{red} = f(Re)$, are presented for an interior (fifth) row and for the first row are represented by Figs. 7.1 through 7.12. The relationships for the interior row of bundles 1 through 12 are listed in Table 4.

Preliminary analysis of the data shown in Figs. 7.1 through 7.12 shows that the heat transfer coefficient increases with row number, attaining a constant value after the third–fourth row. The heat transfer coefficient for the first row is, on the average, 30% lower than that for an interior row. The heat transfer coefficient increases with Re, but no transition to developed turbulent flow occurs over the entire range of Re study, with the exception of bundles 9 and 12. In contrast, the heat transfer coefficient for bundles of bare tubes increases steeply at $Re \simeq 2 \cdot 10^5$, which is seen from the rise in the value of the exponent on Re [67].

It is seen by comparing experimental data on the heat transfer coefficient

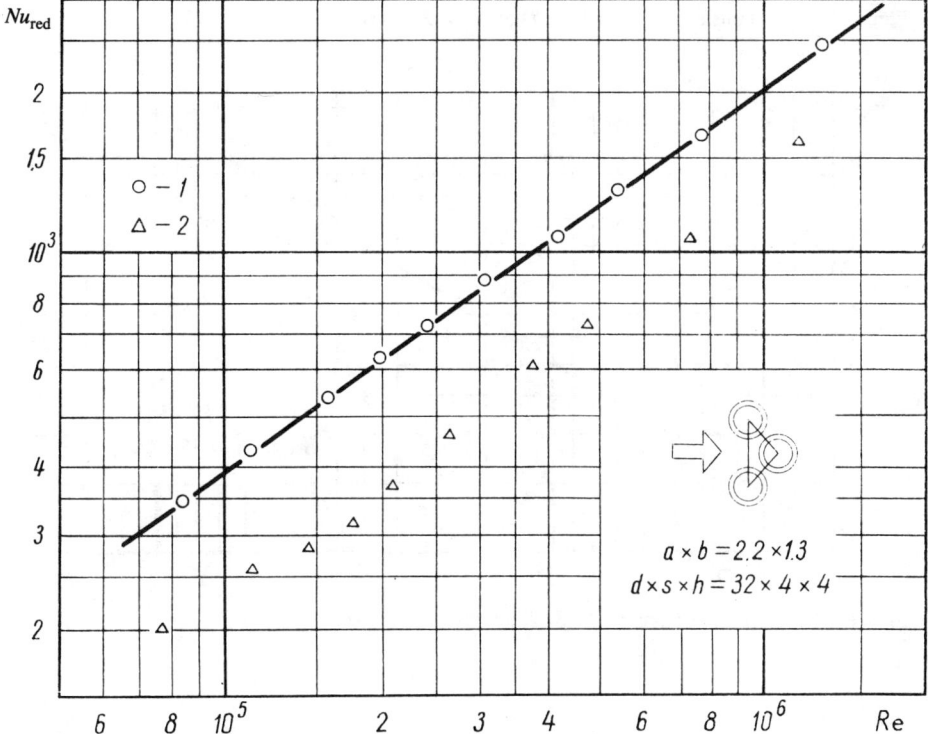

Figure 7.1 Heat transfer from bundle 1. *1)* 5th row; *2)* 1st row.

120 HEAT TRANSFER OF FINNED TUBE BUNDLES IN CROSSFLOW

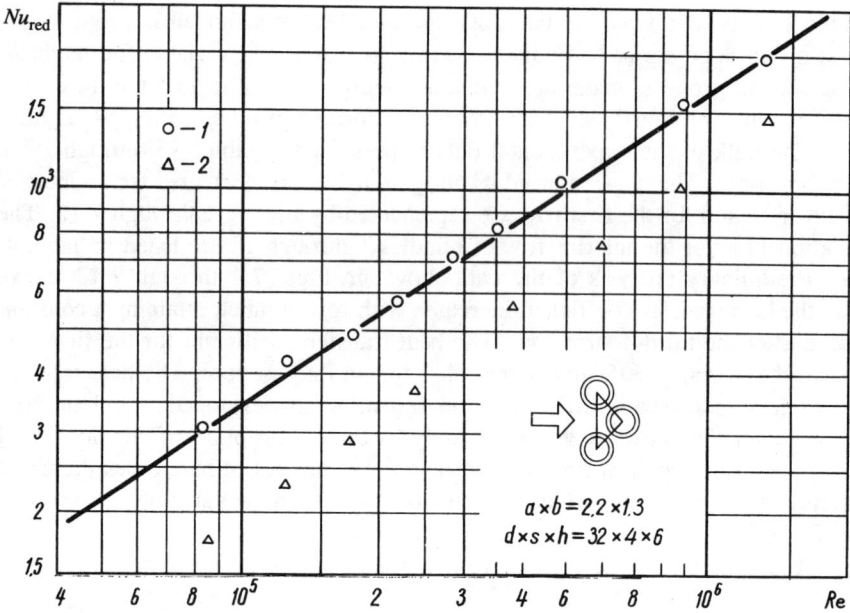

Figure 7.2 Heat transfer from bundle 2. *1)* 5th row; *2)* 1st row.

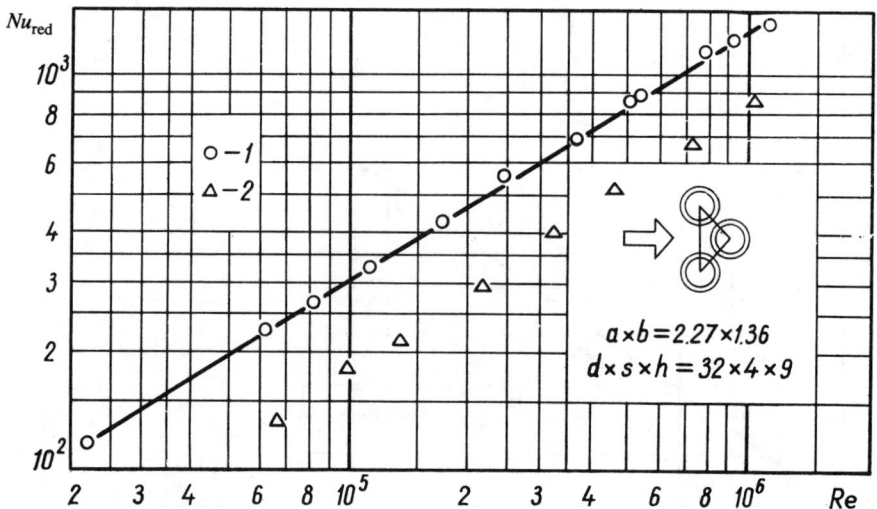

Figure 7.3 Heat transfer from bundle 3. *1)* 5th row; *2)* 1st row.

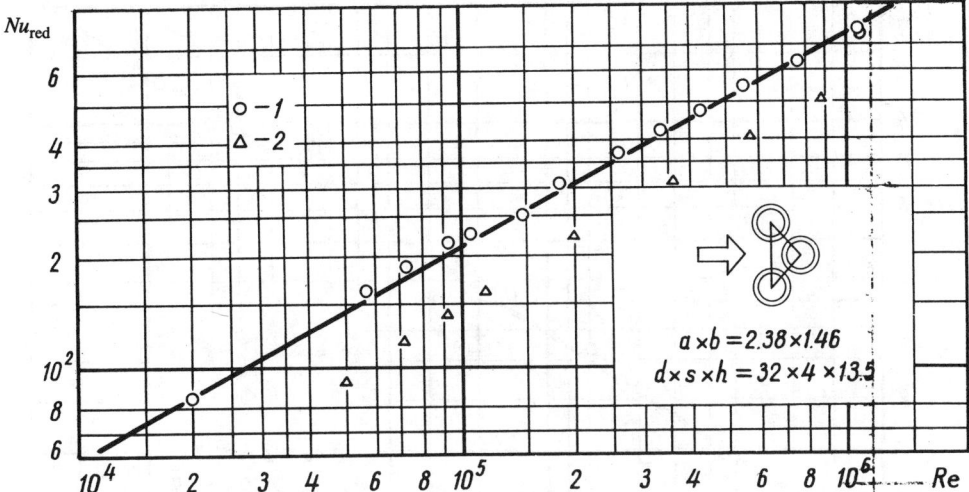

Figure 7.4 Heat transfer from bundle 4. *1)* 5th row; *2)* 1st row.

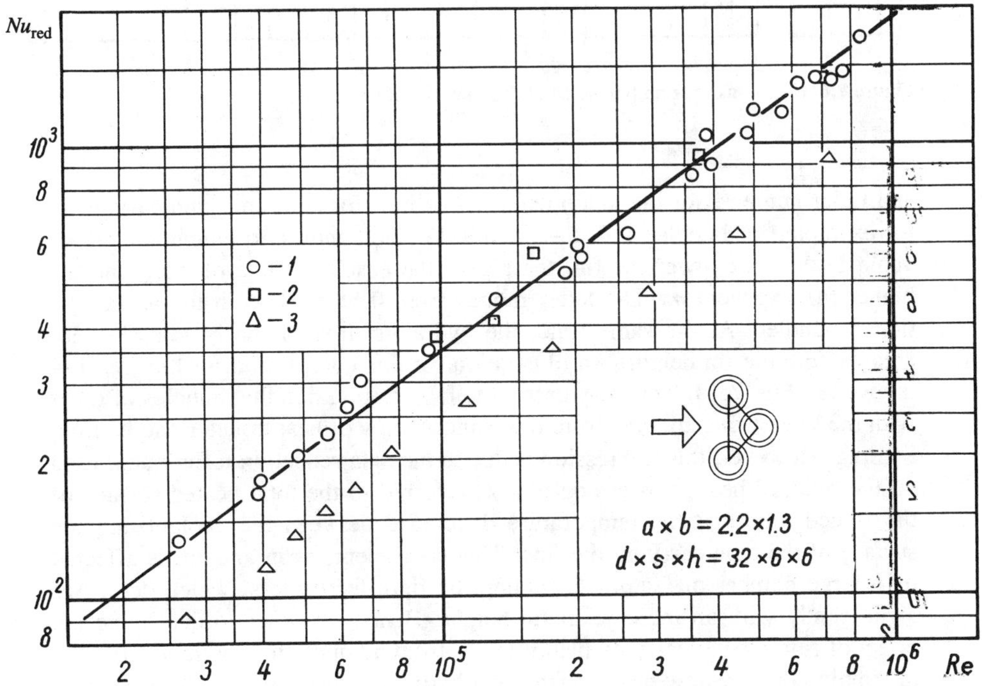

Figure 7.5 Heat transfer from bundle 5. *1)* 5th row; *2)* 1st row; *3)* 1st row.

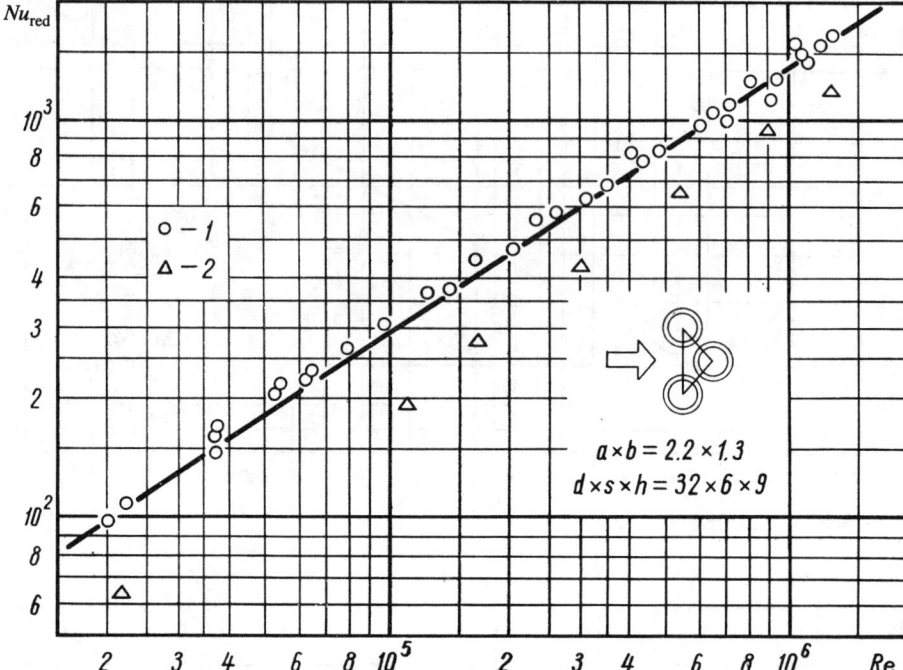

Figure 7.6 Heat transfer from bundle 6. *1)* 5th row; *2)* 1st row.

and Euler numbers for the individual groups (i.e., for constant s and variable m for example, for bundles 1–4, 5–7, and 9–11 (the bundles are numbered as per Table 1) that, the lower the fin, the higher the absolute value of Nu_{red} and the higher the exponent m. The latter ranges from 0.55 to 0.84 in the aforementioned bundles. At the same time, the hydraulic drag of the bundle (Eu) increases with the fin height. An illustration for this comparison for bundles 1–4 is given in Fig. 7.13. This may create the false impression that bundles of tubes with the lowest fins are best from the point of view of heat transfer and hydraulic drag. However, this impression is due to the data reduction technique, based on the reduced heat transfer coefficient, referred to the total heated surface of the finned tube and the temperature difference between the coolant and the surface of the tube carrying the fins. This coefficient, as in known, is affected by a large number of factors, including the fin effectiveness, which decreases significantly with an increase in fin height. Hence, in order to determine the effect of finning geometry on heat transfer from a bundle it is quite insufficient to simply compare raw heat transfer and hydraulic drag data.

MEAN HEAT TRANSFER COEFFICIENT FROM A FINNED TUBE IN A BUNDLE 123

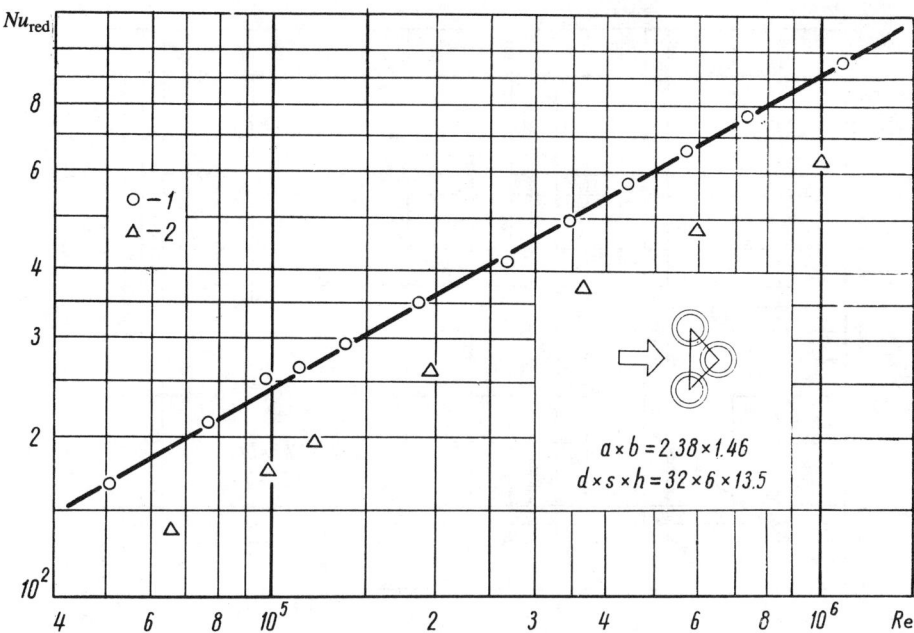

Figure 7.7 Heat transfer from bundle 7. *1)* 5th row; *2)* 1st row.

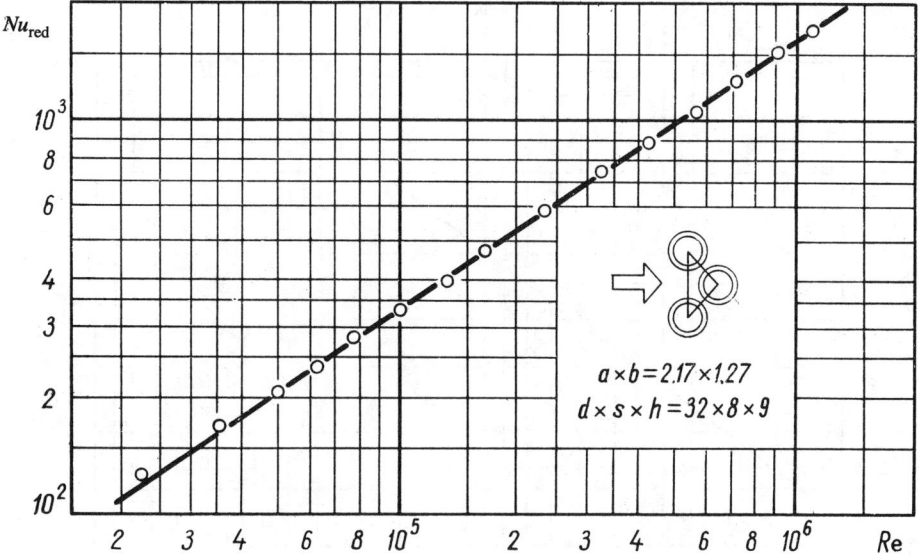

Figure 7.8 Heat transfer from bundle 8.

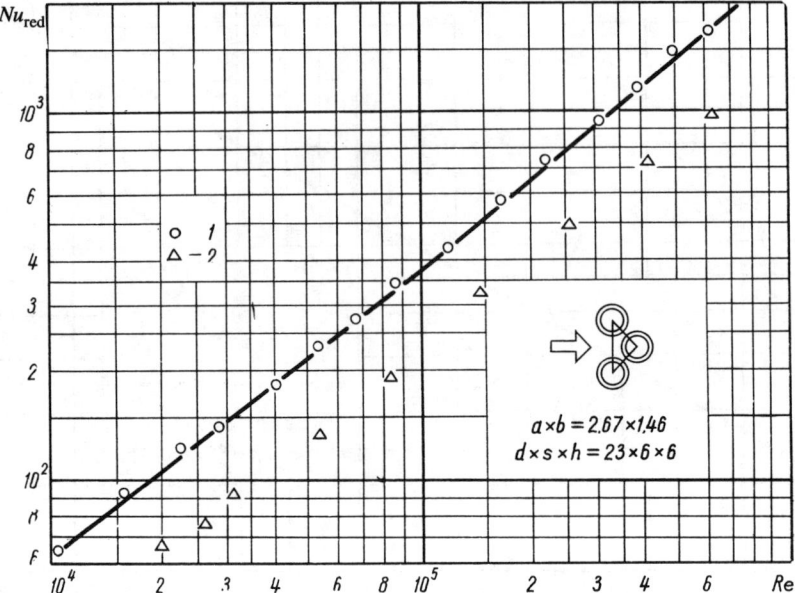

Figure 7.9 Heat transfer from bundle 9. *1)* 5th row; *2)* 1st row.

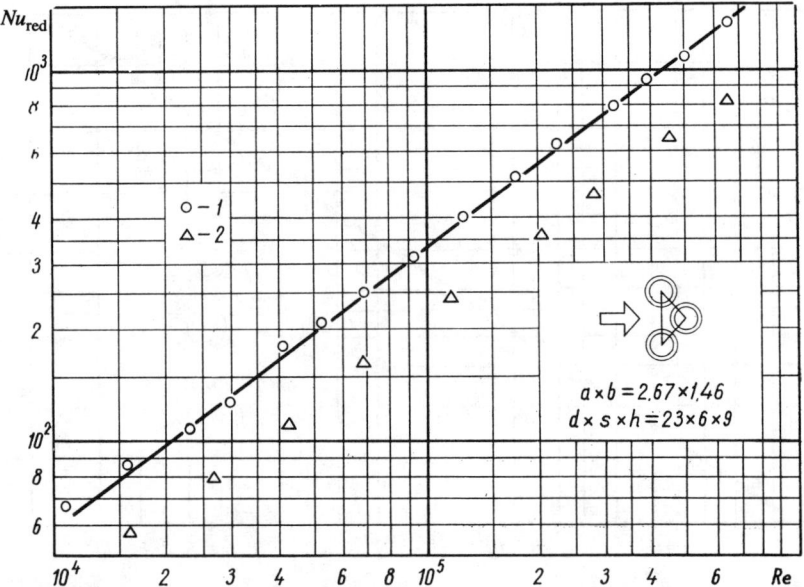

Figure 7.10 Heat transfer from bundle 10. *1)* 5th row; *2)* 1st row.

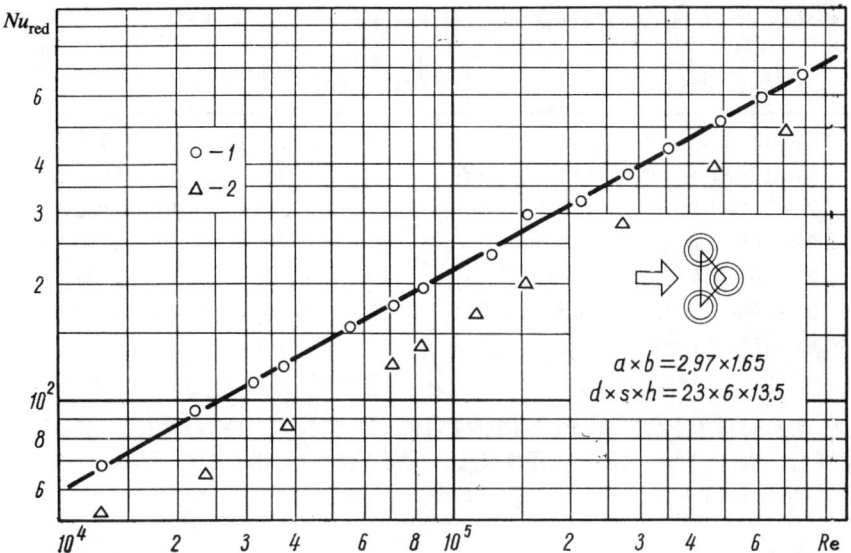

Figure 7.11 Heat transfer from bundle 11. *1)* 5th row; *2)* 1st row.

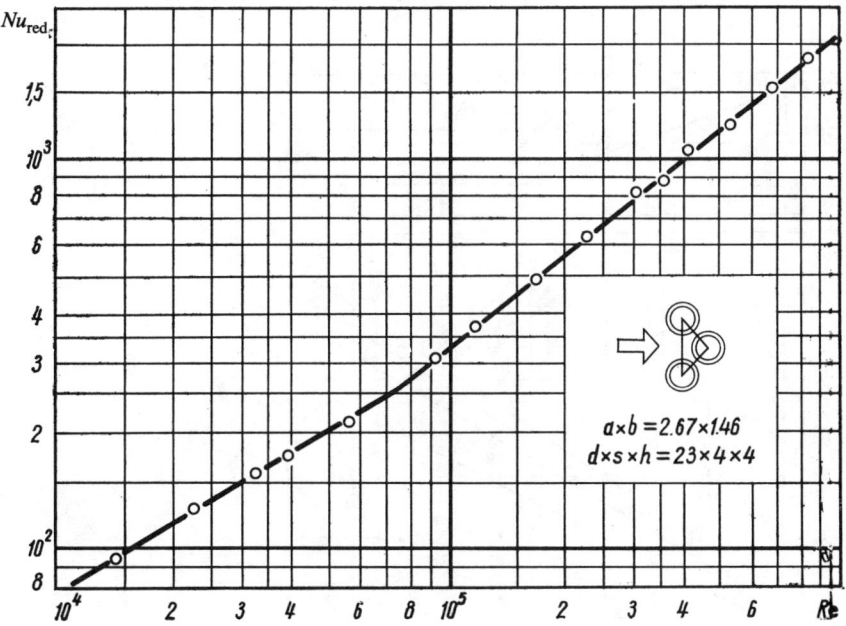

Figure 7.12 Heat transfer from bundle 12.

126 HEAT TRANSFER OF FINNED TUBE BUNDLES IN CROSSFLOW

Table 4 Value of constant c and exponent m in the expression $Nu_{red} = cRe^m$

Bundle No.	Re Range	c	m	Bundle No.	Re Range	c	m
1	$8.5 \cdot 10^4 - 1.3 \cdot 10^6$	0.098	0.72	8	$2.2 \cdot 10^4 - 1.2 \cdot 10^6$	0.124	0.68
2	$8.5 \cdot 10^4 - 1.3 \cdot 10^6$	0.122	0.69	9	$10^4 - 1.2 \cdot 10^5$	0.0525	0.77
3	$2.2 \cdot 10^4 - 1.1 \cdot 10^6$	0.19	0.64		$1.2 \cdot 10^5 - 6.2 \cdot 10^5$	0.022	0.84
4	$2.0 \cdot 10^4 - 1.1 \cdot 10^6$	0.38	0.55	10	$10^4 - 6.5 \cdot 10^5$	0.058	0.75
5	$2.6 \cdot 10^4 - 8.5 \cdot 10^5$	0.069	0.74	11	$1.3 \cdot 10^4 - 8.0 \cdot 10^5$	0.341	0.56
6	$2.2 \cdot 10^4 - 1.2 \cdot 10^6$	0.15	0.66	12	$1.4 \cdot 10^4 - 7.0 \cdot 10^4$	0.335	0.59
7	$5.0 \cdot 10^4 - 1.1 \cdot 10^6$	0.305	0.58		$7.0 \cdot 10^4 - 1.0 \cdot 10^6$	0.026	0.82

7.2. RESULTS OF AN EXPERIMENTAL STUDY OF BUNDLES WITH VARYING GEOMETRIES WITH CONSTANT FIN GEOMETRY

This study was performed with 9 staggered finned tube bundles with different tube arrangements within the bundle. Spiral finned tubes of given geometry ($d \times s \times h \; \delta_2/\delta_1 = 23 \times 6.5 \times 10 \times 2/1$ mm) we used for all the series tests. As shown in Table 2, the relative transverse pitch a of the tubes within the

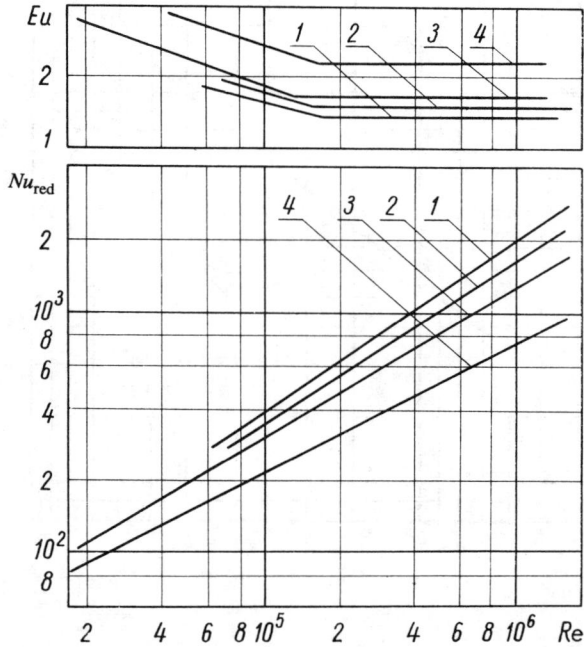

Figure 7.13 Comparison of heat transfer and hydraulic drag of bundles 1 through 4.

bundle ranged from 2.67 to 4.13, and the relative longitudinal pitch b ranged from 1.46 to 2.14. The data obtained in these experiments are listed in the appendices, tables 20 through 27, and are correlated in nondimensional form for the case when the calorimeter tube was placed in the interior (fifth) row in Figs. 7.14 through 7.22. The nondimensional equations obtained for bundles 13 through 21 are listed in Table 5.

It was noted in analyzing the experimental data presented in the figures and tables, that the heat transfer coefficient from the bundles changes only slightly for the range of a and b under study—only by 20%, whereas the Euler number changes by up to 60%.

The effect of the transverse and longitudinal bundle pitches on heat transfer is clearly seen in Fig. 7.23, which compares data for the three most different bundle configurations (bundles 18, 20, and 21). Analysis shows that the heat transfer coefficient increases insignificantly ($\sim 3\%$) with an increase in a, whereas a reduction in b results in a significant ($\sim 20\%$) rise (bundles 18 and 21). This result is somewhat unexpected, since according to Zozulya, et. al., [34] and Yudin with Tokhtarova [68], heat transfer is much more affected by changes in a.

To clarify this situation, we compared the range of ratios of relative pitches used in our studies and in those of references [34] and [68]. This is done by Fig. 7.24, from where it is seen that the range of values of b of these studies overlaps, whereas the values of a in our experiments are higher.

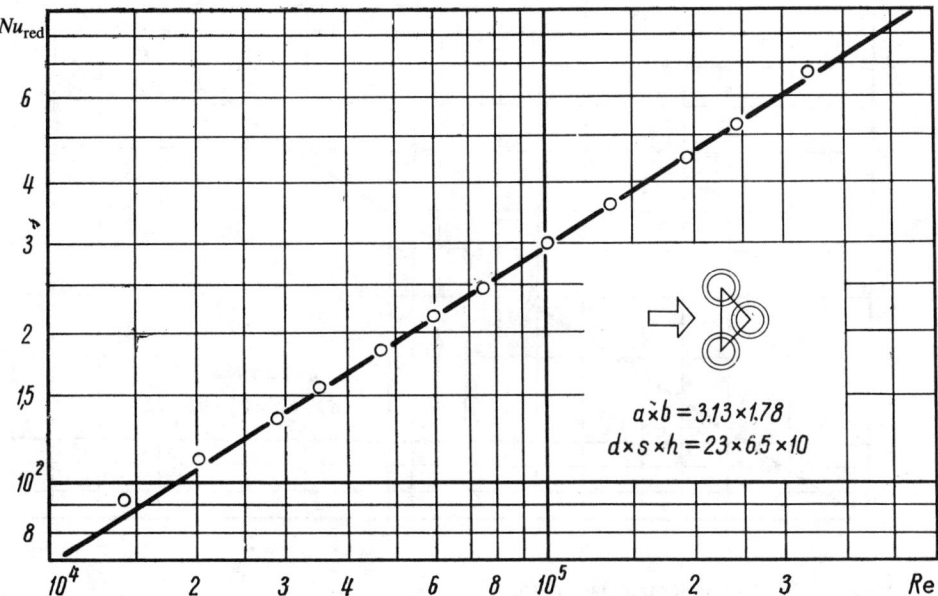

Figure 7.14 Heat transfer from bundle 13.

128 HEAT TRANSFER OF FINNED TUBE BUNDLES IN CROSSFLOW

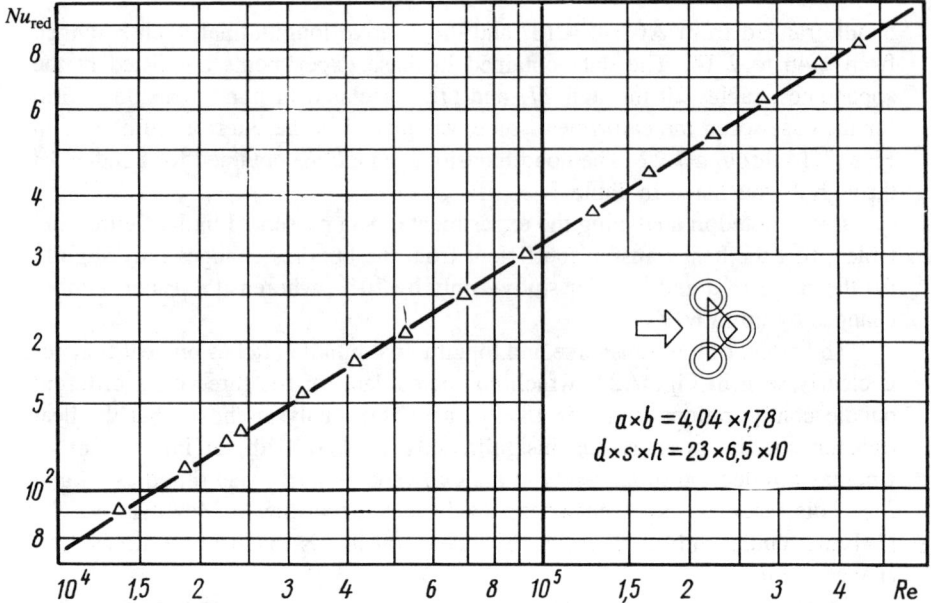

Figure 7.15 Heat transfer from bundle 14.

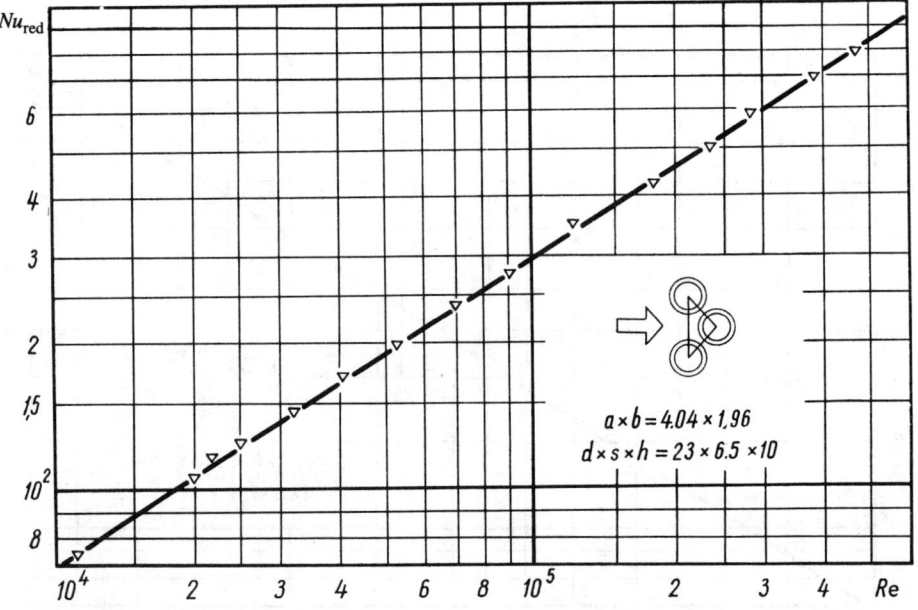

Figure 7.16 Heat transfer from bundle 15.

MEAN HEAT TRANSFER COEFFICIENT FROM A FINNED TUBE IN A BUNDLE

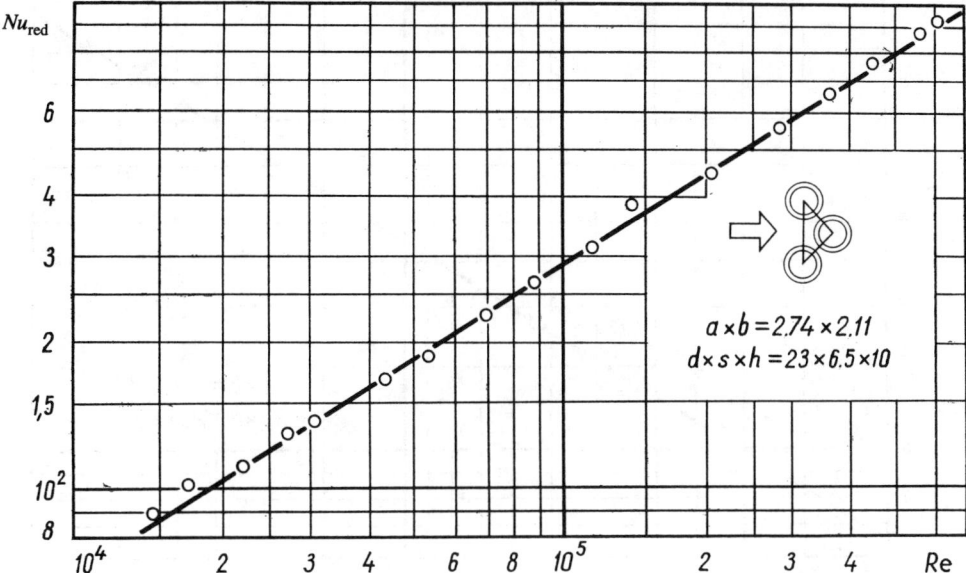

Figure 7.17 Heat transfer from bundle 16.

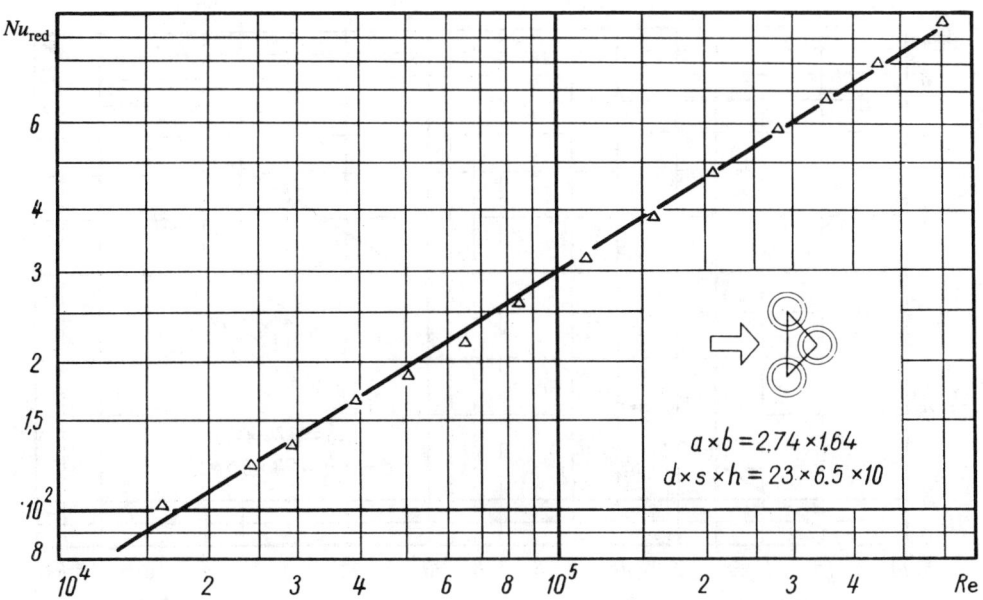

Figure 7.18 Heat transfer from bundle 17.

130 HEAT TRANSFER OF FINNED TUBE BUNDLES IN CROSSFLOW

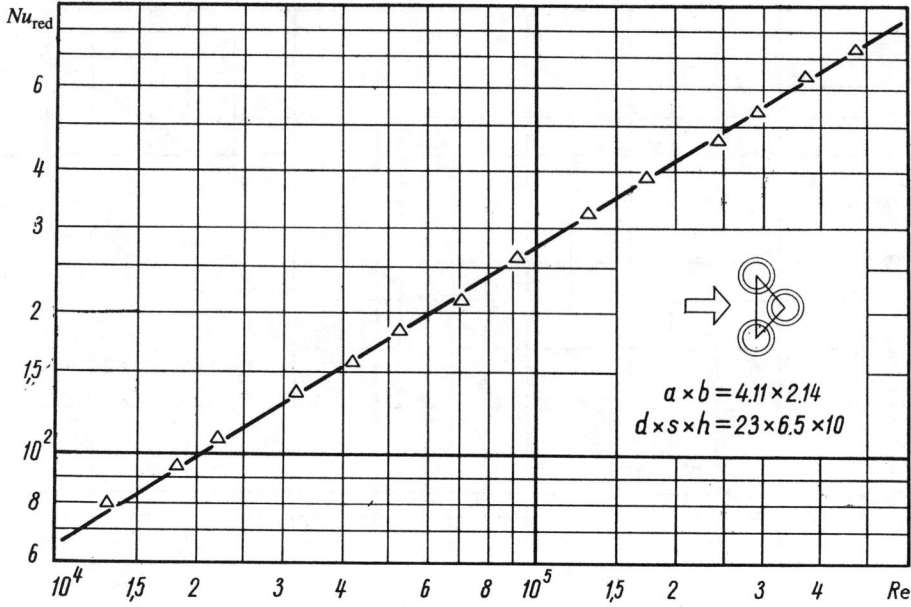

Figure 7.19 Heat transfer from bundle 18.

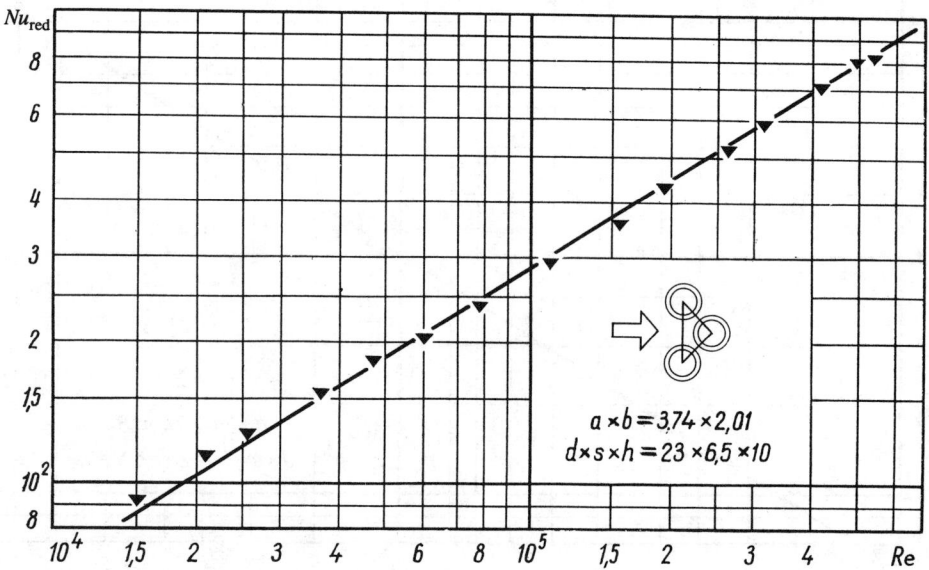

Figure 7.20 Heat transfer from bundle 19.

MEAN HEAT TRANSFER COEFFICIENT FROM A FINNED TUBE IN A BUNDLE 131

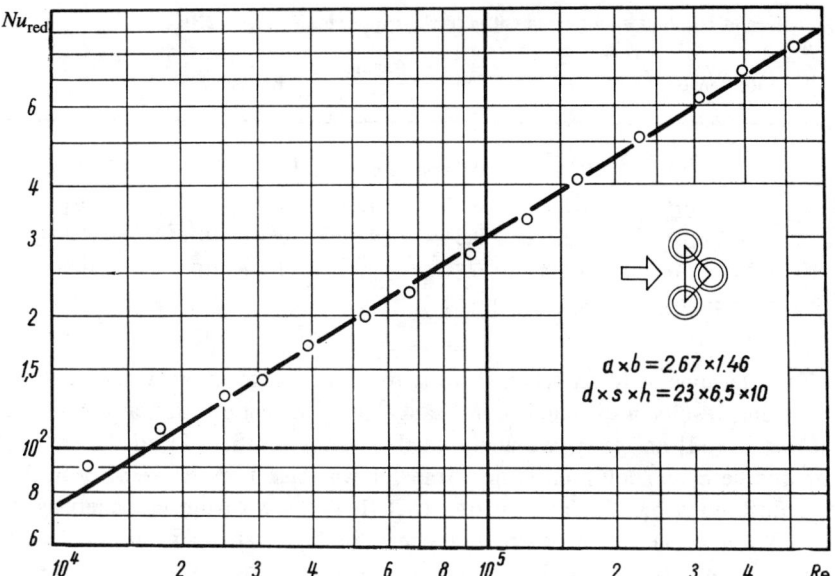

Figure 7.21 Heat transfer from bundle 20.

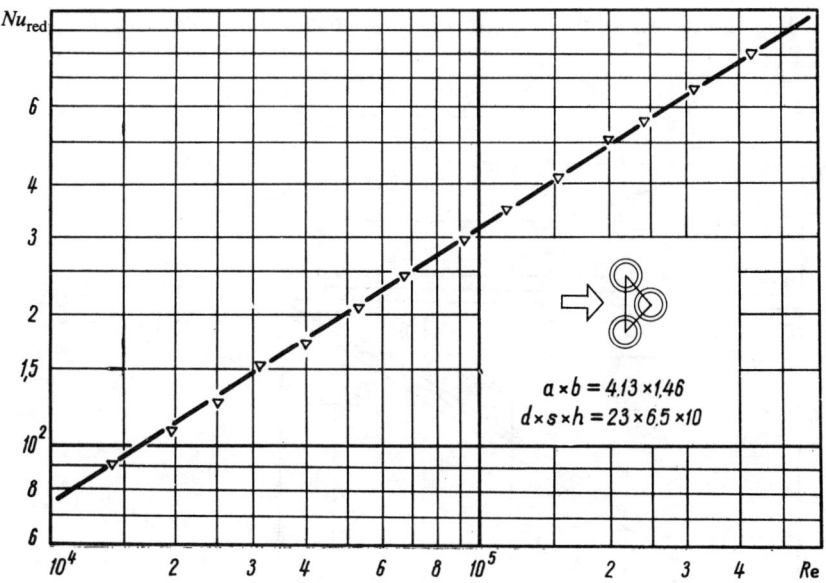

Figure 7.22 Heat transfer from bundle 21.

132 HEAT TRANSFER OF FINNED TUBE BUNDLES IN CROSSFLOW

Table 5 Values of constant c and exponent m in the expression $Nu_{red} = cRe^m$

Bundle No.	Re Range	c	m	Bundle No.	Re Range	c	m
13	$10^4 - 5 \cdot 10^5$	0,19	0.64	18	$10^4 - 5 \cdot 10^5$	0.17	0.63
14	$10^4 - 5 \cdot 10^5$	0,202	0.64	19	$10^4 - 6 \cdot 10^5$	0.184	0.64
15	$10^4 - 5 \cdot 10^5$	0,19	0.64	20	$10^4 - 5 \cdot 10^5$	0.273	0.61
16	$10^4 - 6 \cdot 10^5$	0,184	0.64	21	$10^4 - 5 \cdot 10^5$	0.198	0.63
17	$10^4 - 6 \cdot 10^5$	0,19	0.64				

To determine why an increase in a over the range of its values in our experiments results in such an insignificant change in heat transfer, as compared with that in [34] and [68], we analyzed the flow velocities. In all the studies compared here, including ours, heat transfer was based on the velocity in the most constrained bundle cross section (w_y). It is known from our analysis in Sec. 4.3 (Fig. 4.7) that in order to compare the heat transfer from bundles of various configurations with a high accuracy, it is best to do so on the basis of the mean velocity w_m, since a change in a does not always result in a similar change in w_m/w_y. In denser bundles, when $a < 2.5$, w_m is closer to the free-stream velocity, and correlating heat transfer in terms of w_y results in significant error. The range of variation of a in [86] was from 1.7 to 2. In this case, according to Fig. 4.7, the change in w_m/w_y is greater than 60%.

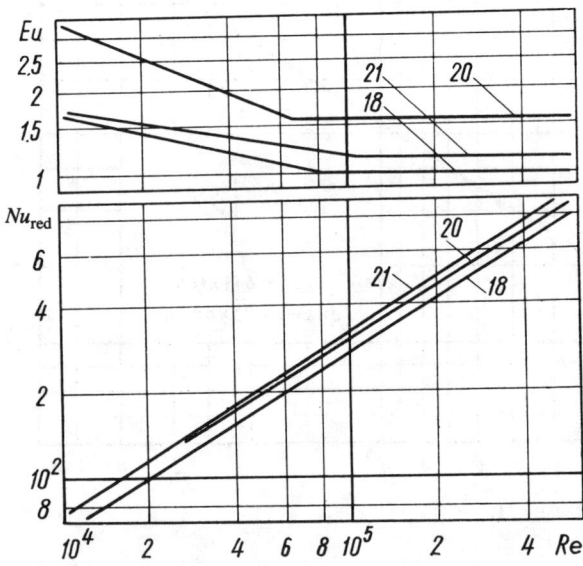

Figure 7.23 Comparison of heat transfer and drag of bundles 18, 20 and 21.

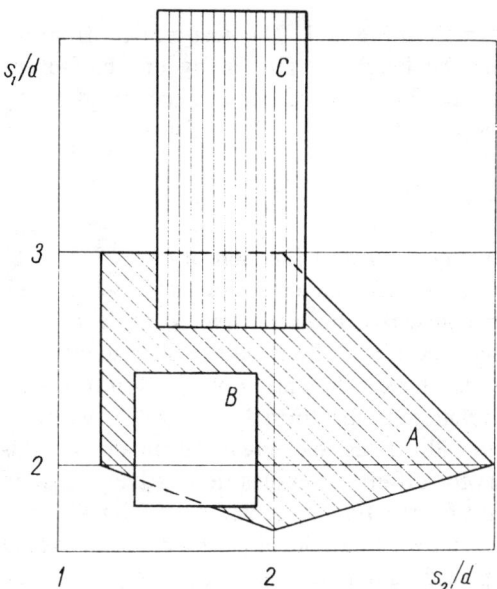

Figure 7.24 Comparison of ranges of relative pitches of finned tube bundles in the experiments by various investigators. *A)* According to Yudin with Tokhtarova [68]; *B)* according to Zozulya, et al. [34]; *C)* present study.

The value of a in our experiments ranged from 2.67 to 4.13, and the change in w_m/w_y was only 9%. It follows that, at moderate a, an increase in s_1 should yield a higher gain in heat transfer, which was in fact obtained in [34] and [68]. At higher a, heat transfer from a bundle depends more on s_2 than on s_1.

7.3. CORRELATION OF EXPERIMENTAL DATA

Preliminary reduction of experimental results on the basis of the reduced heat transfer coefficient (see tables 4 and 5) is, due to the limited nature of the reduction technique, suitable only for describing data obtained for the bundle in question. A more extensive correlation can be obtained only when the experimental data are reduced on the basis of the mean convective heat transfer coefficient α. The latter is based on the mean temperature of the entire finned surface.

Since mean temperature data are not available, the convective heat transfer coefficients were obtained analytically, on the basis of results presented in the preceding chapter. A curve of $\alpha_{red} = f(\alpha)$ was constructed for each bundle on the basis of Eq. (6.11).

It was found by comparing the effectiveness of spiral fins (calculated from Eq. (2.71) and circumferential fins (calculated from Fig. 2.7) that, within the limits of the fin angle of attack assumed in our study, they differ by not more than 0.5%. Since no nomograms are available for determining E for spiral

134 HEAT TRANSFER OF FINNED TUBE BUNDLES IN CROSSFLOW

fins, we used for convenience the graph in Fig. 2.7. The heat transfer from the fin tip was included by increasing the height of the fin by one half of its thickness. The trapezoidal shape of the fin cross section was accounted for by using coefficient ξ, which was determined from Fig. 3.3 as a function of βh and

$$\sqrt{\delta_2/\delta_1}$$

Coefficient ψ was determined from Fig. 6.11 as a function of βh.

Fig. 7.25 shows the curve of $\alpha_{red} = f(\alpha)$, calculated for bundles 1 through 4, from which the corresponding convective heat transfer coefficients were determined. Such graphs were calculated for all the bundles under study.

The values of convective heat transfer coefficients α obtained in this manner was used for obtaining a correlation in the form Nu = f(Re), which is plotted in Fig. 7.26. This figure illustrates the great advantage of this correlation—the experimental points for different bundles lie relatively close to one another even without correction for the finning geometry and bundle configuration (the maximum difference between individual points does not exceed 40%). The points tend to stratify upward with increasing Re, which points to an increase in the heat transfer rate, occurring upon transitions to predominantly turbulent flow. Subsequent data reduction was performed with correction for the finning geometry, represented by parameters s/d and h/d, which are the most widely used and the most convenient for calculations. The fin thickness was not included among the finning parameters due to its insignificance. The bundle configuration was accounted for by the ratio a/b. This means that the

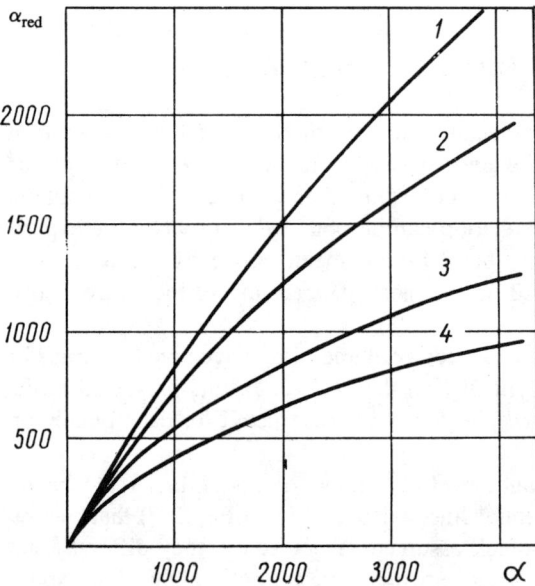

Figure 7.25 Graph for determining convective heat transfer coefficients of bundles 1 through 4.

MEAN HEAT TRANSFER COEFFICIENT FROM A FINNED TUBE IN A BUNDLE 135

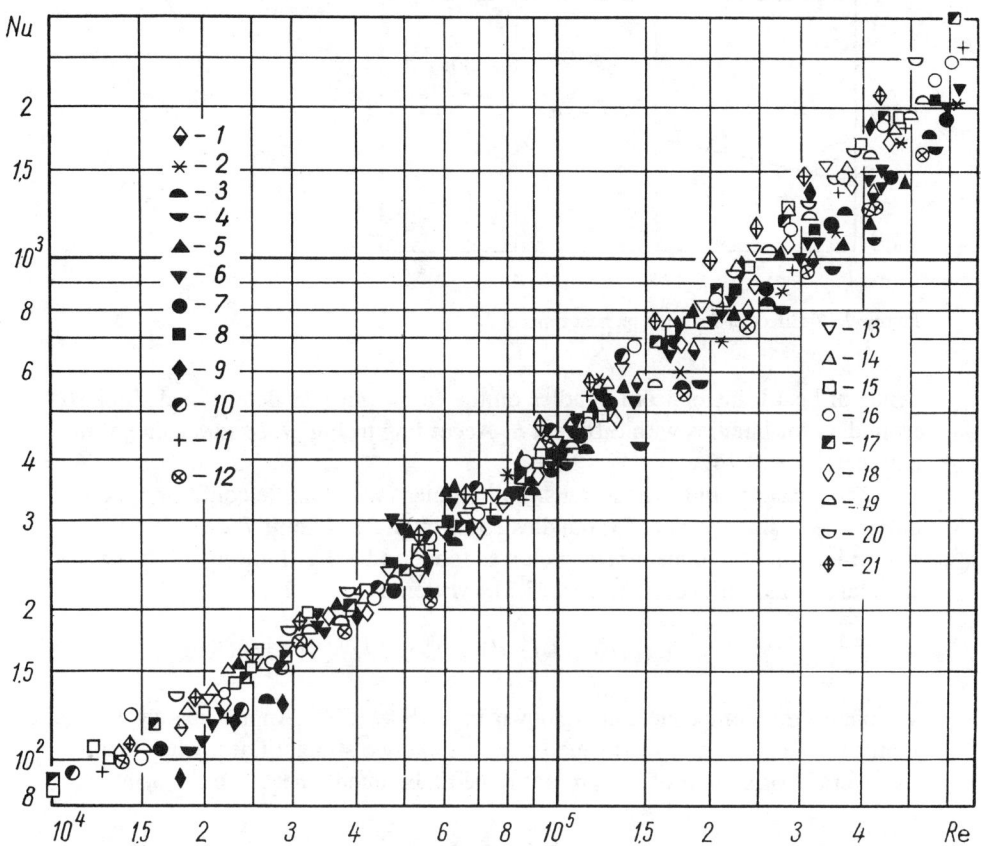

Figure 7.26 Experimental data on heat transfer from bundles 1 through 21, reduced on the basis of the convective heat transfer coefficients as Nu = f(Re).

experimental points for bundles 1 through 21 were correlated with the expression

$$\mathrm{Nu} = c\,(s/d)^k\,(h/d)^l\,(a/b)^m\,\mathrm{Re}^n \qquad (7.2)$$

Since heat transfer from bundles of tubes with different fin pitches (bundles 3, 6 and 8), expressed in the form Nu = $c_i \mathrm{RE}^n$, is different (see Fig. 7.26), then, naturally, constant c_i changes from one bundle to another. A plot of $c_i = f(s/d)$ is given in Fig. 7.27. It is expressed as

$$c_i = c_1\,(s/d)^{0.18} \qquad (7.3)$$

This means that the value of exponent k in Equation (7.2) is 0.18. The depen-

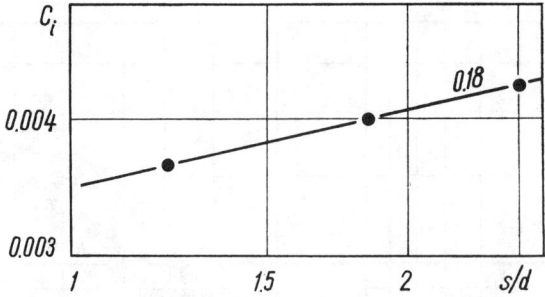

Figure 7.27 Effect of fin pitch on heat transfer.

dence of heat transfer from bundles on the fin height was determined similarly from data for bundles with different h. According to Fig. 7.28, the value of this exponent is -0.14.

The variation in the heat transfer coefficient with bundle configuration was determined using the data for bundles 13 through 21. Figure 7.29 illustrates the determination of exponent m, which was found to be 0.2. Plotting all the experimental points, reduced with Eq. (7.2), written as

$$(a/b)^{-0.2} (s/d)^{-0.18} (h/d)^{0.14} \mathrm{Nu} = f(\mathrm{Re})$$

we obtain the composite graph shown in in Fig. 7.30. Approximating all the points by the method of least squares, we obtain two straight lines, which express the overall correlation of the present finned tube bundle heat transfer data.

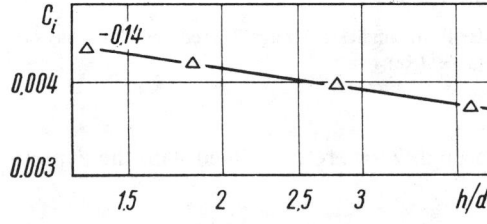

Figure 7.28 Effect of fin height on heat transfer.

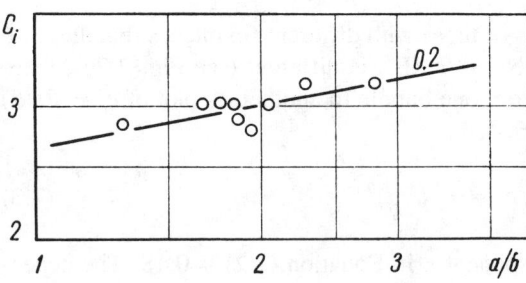

Figure 7.29 Effect of bundle configuration on heat transfer.

MEAN HEAT TRANSFER COEFFICIENT FROM A FINNED TUBE IN A BUNDLE

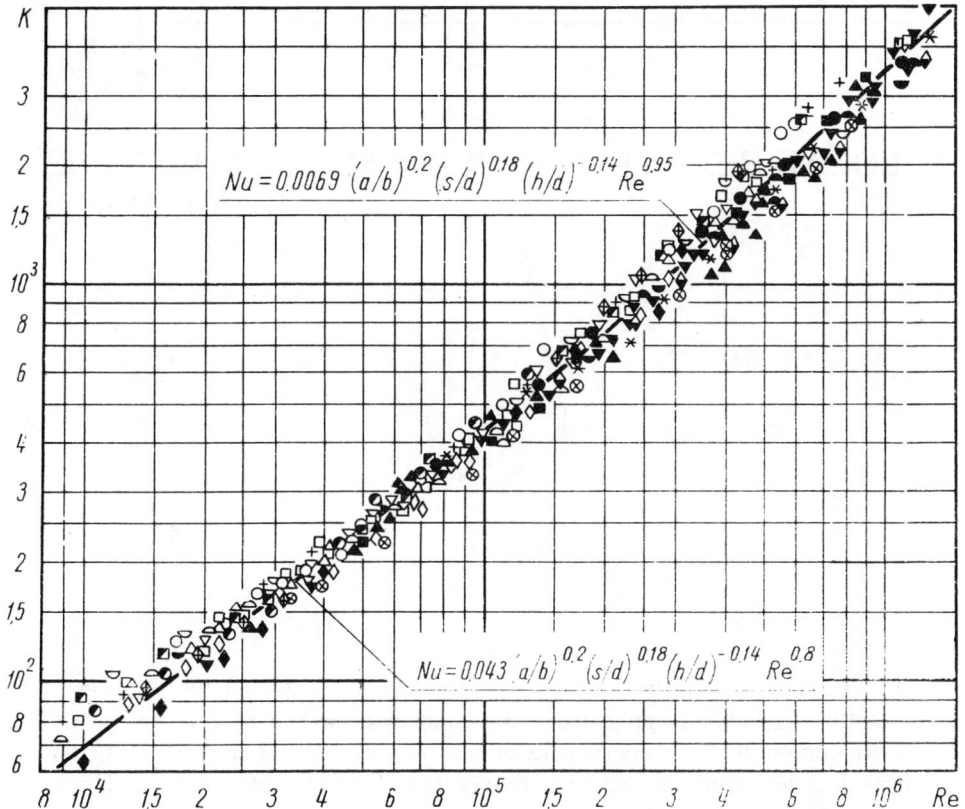

Figure 7.30 Correlation of heat transfer data for bundles 1 through 21.

The final form of (7.2) for Re from 2×10^4 to 2×10^5 becomes

$$\mathrm{Nu} = 0.044 \, (a/b)^{0.2} \, (s/d)^{0.18} \, (h/d)^{-0.14} \, \mathrm{Re}^{0.8} \qquad (7.4)$$

and for Re from 2×10^5 to 1.3×10^6

$$\mathrm{Nu} = 0.0067 \, (a/b)^{0.2} \, (s/d)^{0.18} \, (h/d)^{-0.14} \, \mathrm{Re}^{0.95} \qquad (7.5)$$

It is seen from Fig. 7.30 that the experimental data for bundles 1 through 21 are described by Eqs. (7.4) and (7.5) with a scatter of $\pm 14\%$. Note that equation 7.4 and 7.5 are for *air flows only*. For other fluids, the Pr has to be taken into account (see equations 10.11 and 10.12).

CHAPTER
EIGHT

LOCAL HEAT TRANSFER FROM A FINNED TUBE IN A BUNDLE

It was noted previously that the flow pattern over a finned tube bundle in crossflow is rather complicated. This complexity should be reflected in the distribution of the local coefficients of heat transfer over the surface of a finned tube. It is seen from the results of Lymer with Ridal [51] and Neal with Hitchcock [57] that, due to the complex geometry of finned tubes, flow over them causes the local heat transfer coefficients to vary both circumferentially and radially. This variation is also a function of the finning geometry, the arrangement of tubes within the bundle, and the flow mode. Hence experimental study of the distribution of the local heat transfer coefficient is important in investigating the mechanism of heat transfer, primarily in the case of predominantly turbulent flows. Below, we present experimental data obtained for the local heat transfer coefficient for a finned tube in the range of Re from $4.8 \cdot 10^4$ to $7 \cdot 6 \cdot 10^5$.

8.1. PRELIMINARY REMARKS

Simulation of local heat transfer from a finned tube is a rather difficult task. Two simulation methods may be used, namely a method based on heating the entire finned surface, and, secondly, a method of local (point) heating, when only a small part of the fin is heated. In the latter method the experimental tube is rotated about its axis, and the heated part traverses the entire flow over the fin. The local heat transfer is determined by measuring the surface temperature of the segment and the heat flux density.

Using the first method, the conditions obtained at the model are rather close to those occurring on a real tube. Under the second method the conditions at the model evidently do not correspond to those actually encountered. The reason for this is the delay in the development of the thermal boundary layer above the heated part as compared with that of the hydrodynamic boundary layer. In spite of this, such a method makes possible to obtain the local heat transfer pattern for the case when basically only the hydrodynamic boundary layer exists at the surface, which is quite interesting because the flow pattern is the basic factor controlling the rate of heat transfer from a tube bundle.

The local heat transfer pattern which is obtained using the local heating method more accurately reflects the distribution of velocity and flow turbulence at the fin surface. This method was, therefore, used in the present studies. The experimental tube is depicted schematically in Fig. 4.6.

When investigating the local heat transfer coefficient variation for a finned tube in cross flow, it is useful to consider for comparison the distribution of the coefficient over the surface of a bare tube.

The circumferential distribution of the local heat transfer coefficient from a circular tube to an air flow is shown as a function of Re in Fig. 8.1, replotted

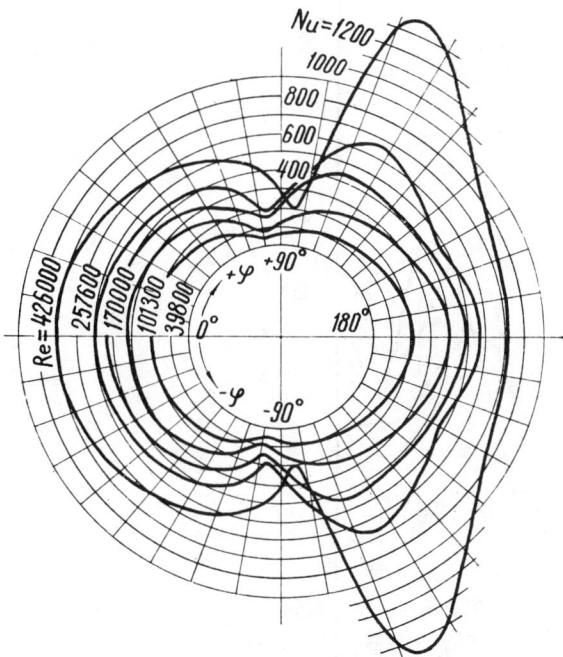

Figure 8.1 Local heat transfer coefficients to a circular cylinder placed in cross flow of air. *1)* Re = 3.9 × 10⁴; *2)* 1.01 × 10⁵; *3)* 1.7 × 10⁵; *4)* 2.5 × 10⁵; *5)* 4.2 × 10⁵.

from the book by Gröber, et. al. [88]. The distribution of heat transfer over the cylinder surface is rather complex, varying significantly as a function of the flow pattern. It is seen that a rise in Re is accompanied not only by an increase in the overall level of heat transfer coefficient from the tube, but also by a modification of the distribution of the heat transfer coefficient over the cylinder's circumference. These changes occur primarily in the trailing part of the cylinder. The nature of the change in the heat transfer coefficient for the leading part (up to the point of separation, $\varphi \simeq 90°$) is similar at all Re. At high Re ($> 10^5$), the laminar boundary layer forming a leading part of the cylinder is turbulized, becomes turbulent, and finally breaks off, the breakoff point shifting downstream (Žukauskas [89]) to $\varphi = 140$–$150°$. These phenomena are reflected in the heat transfer coefficient distribution by the appearance of two minima—the first pertaining to the start of transition flow, and the second to the separation of the boundary layer (*see* Fig. 8.1). The heat transfer coefficient in the trailing part of the tube increases due to the high vorticity and turbulence. It should be noted that the data described in Fig. 8.1 were determined by the total heating technique. It should also be noted that the distribution of local heat

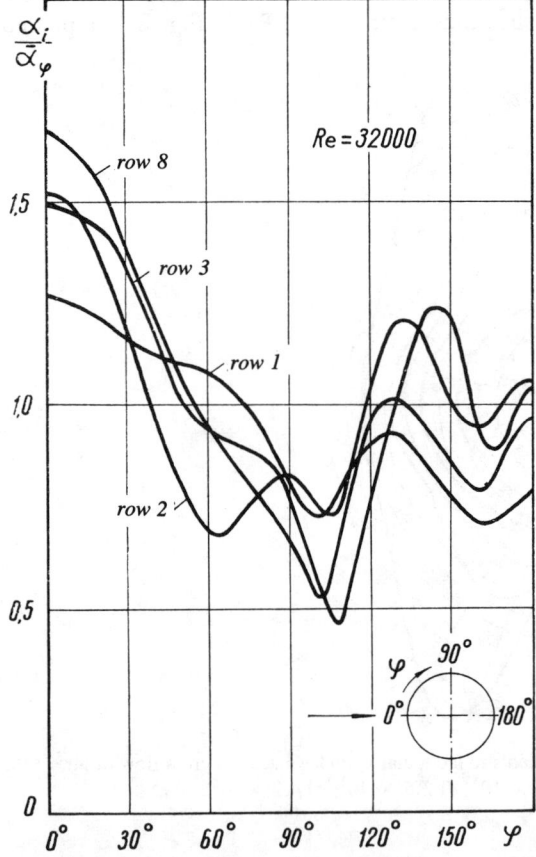

Figure 8.2 Distribution of local heat transfer coefficients over the circumference of a bare tube, placed in different rows of a 1.5 × 1.08 staggered bundle.

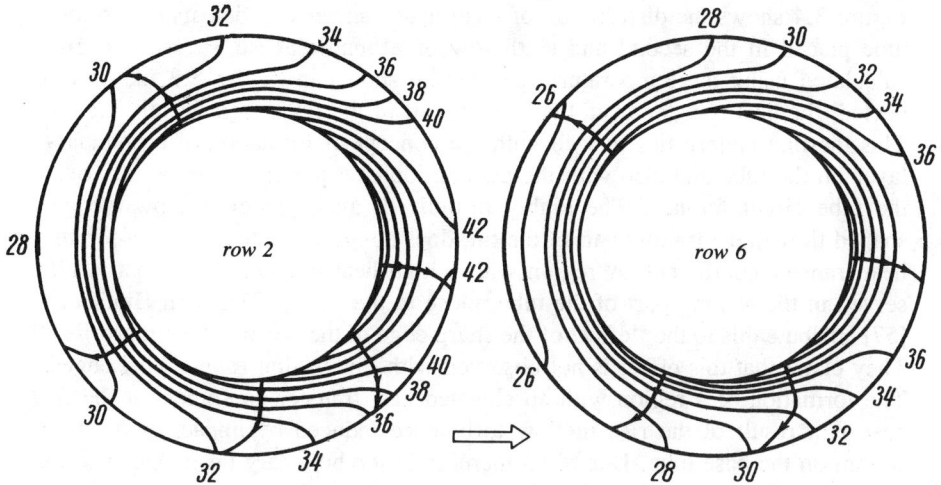

Figure 8.3 Temperature distribution over the surface of a finned tube placed in rows 2 and 6 of a staggered bundle.

transfer over the circumference of the tube changes significantly when such a tube is made a part of a bundle. Figure 8.2 shows the variation in the relative local heat transfer coefficient $\alpha_l/\bar{\alpha}\varphi$ (where $\bar{\alpha}\varphi$ is the heat transfer coefficient averaged over the circumference of the tube) over the circumference over a bare tube, placed in different rows of a staggered bundle at Re = 3.2×10^4 [90]. It is seen that there is a great difference in the distribution of the heat transfer coefficient in the first and interior rows respectively. However, for this range of Re, the maximum of heat transfer coefficient for all the rows occurs at the leading point of the tube. According to the study by Žukauskas, et. al. [91], performed with water, (basically in the supercritical region of Re), this situation does not hold at high values of Re. Depending on the bundle configuration and the value of Re, the maximum heat transfer coefficient from the first row of a staggered bundle occur in the trailing zone of the tube, and the maximum of interior rows is observed at φ from 90 to 110°.

The nature of distribution of the heat transfer coefficient over the surface of a finned tube may be judged from the data obtained by Neal and Hitchcock [57] in experiments with a fully heated model of a finned tube with $d \times s \times h \times \delta = 152 \times 17 \times 40 \times 5$ mm. The model tube was placed in the second and sixth rows of a staggered tube bundle ($a \times b = 2.86 \times 1.37$). Figure 8.3 presents the temperature distribution over the fin surface, obtained with the model. The isotherms show that the temperature of the trailing part of the fin, particularly in the second row, is significantly higher, for which reason heat is observed to flow from this part to the leading part of the fin. Within the leading and the trailing parts of the fin, heat flows from the base to the tip of the fin. The high temperature differences between the tip and base of the fin, on the other hand, indicate that the thermal effectiveness of the fin under study is low.

Figure 8.4 shows the distribution of local heat transfer coefficients of a finned tube placed in the second and sixth row of a bundle at Re = 1.25×10^5, calculated using the temperature distribution shown in Figure 8.3. As is expected, the heat transfer from the leading part of the fin is significantly higher. This thermal pattern ties in well with the concept of formation of a boundary layer on the tube and also with the assumption of separation at $\varphi = 90°$ over the tube circumference. The highly turbulized, and apparently slow wedge-shaped flow in the trailing part of the fin, does not give rise to an increase in the heat transfer coefficient. A region with a high heat transfer coefficient is observed in the leading part of the tube, close to the fin tip. Neal and Hitchcock [57] attribute this to the "effect of the sharp edge of the rectangular fin profile." They claim that this effect is not observed with fins having rounded off edges. The formation of a region with an elevated heat transfer coefficient at the fin base is a result of the rise in flow turbulence induced by impact of the free stream on the base tube. Due to an increase in the boundary layer thickness as the stream flows over the tube, the heat transfer coefficient on the base tube decreases. Further investigations with the same tube models in staggered bundles of different configurations ($a \times b = 2.68 \times 1$ and 2×1.74) showed that the mean heat transfer from the sixth row was significantly less than that of the 2nd row.

Similar studies were performed by Lymer and Ridal [51] on models of a finned tube with $d \times s \times h \times \delta = 133 \times 13 \times 38 \times 3.3$ mm in a staggered bundle ($a \times b = 2 \times 1.35$) at Re = $2-29 \times 10^4$. Heating in this case was by

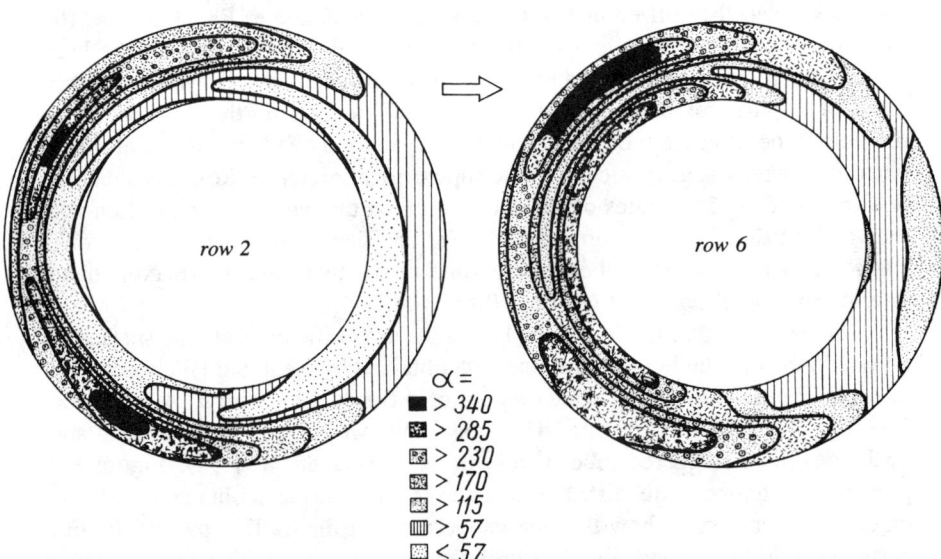

Figure 8.4 Distribution of local heat transfer coefficients at the surface of a finned tube in rows 2 and 6 of staggered bundle.

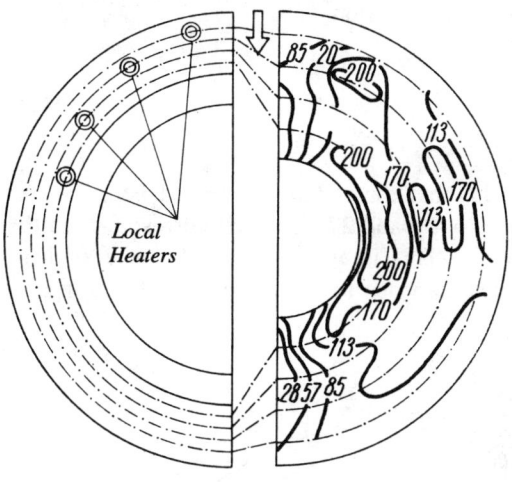

Figure 8.5 Distribution of local heat transfer coefficients on a finned tube placed in row 2 of a staggered bundle.

the point technique. Figure 8.5 presents data on local heat transfer, obtained for a tube in the second row of a bundle placed in a crossflow of air.

It is seen by examining the data plotted in Figs. 8.4 and 8.5 that the heat transfer distribution pattern over the fin surface in both studies was different due to the use of different techniques for heating the model. The maximum heat transfer coefficient for a model with complete heating occurs much closer to the leading point of the finned tube than it does in the point-heated model. As is seen from Figs. 8.1 and 8.5, significant differences are observed in the pattern of the heat transfer coefficient distribution on finned and bare tubes respectively.

8.2. EXPERIMENTAL RESULTS

The investigations were performed with the model shown in Fig. 4.6. The local heat transfer was investigated on the surface of a single tube, placed in a 140 mm wide channel with Re values of 4.8×10^4, 1.6×10^5 and $7 \times 6 \cdot 10^5$. Measurements were also performed in the fifth row of 2.38×1.46 staggered bundle at Re = 1.8×10^5.

1. Variation in a Radially Averaged Heat Transfer Coefficient Around the Circumference of a Finned Tube

The circumferential distribution of the radial average of the local heat transfer coefficients i.e., (averaged over the height of the fin) was obtained by rotating the experimental tube about its axis. Such data were obtained both with a single tube, and with a tube within a bundle, and are shown in Fig. 8.6.

144 HEAT TRANSFER OF FINNED TUBE BUNDLES IN CROSSFLOW

Averaging was performed using the equation

$$\bar{\alpha}_h = \frac{\Sigma \alpha_i F_i}{\Sigma F_i} \tag{8.1}$$

It is seen from the data in Fig. 8.6 that the radially averaged heat transfer coefficients are at maximum near $\varphi = 90$ and $270°$, which corresponds to the narrowest flowpassage in the bundle. This is entirely natural, since at these points the flow velocity is at a maximum due to constriction of the flowpass-

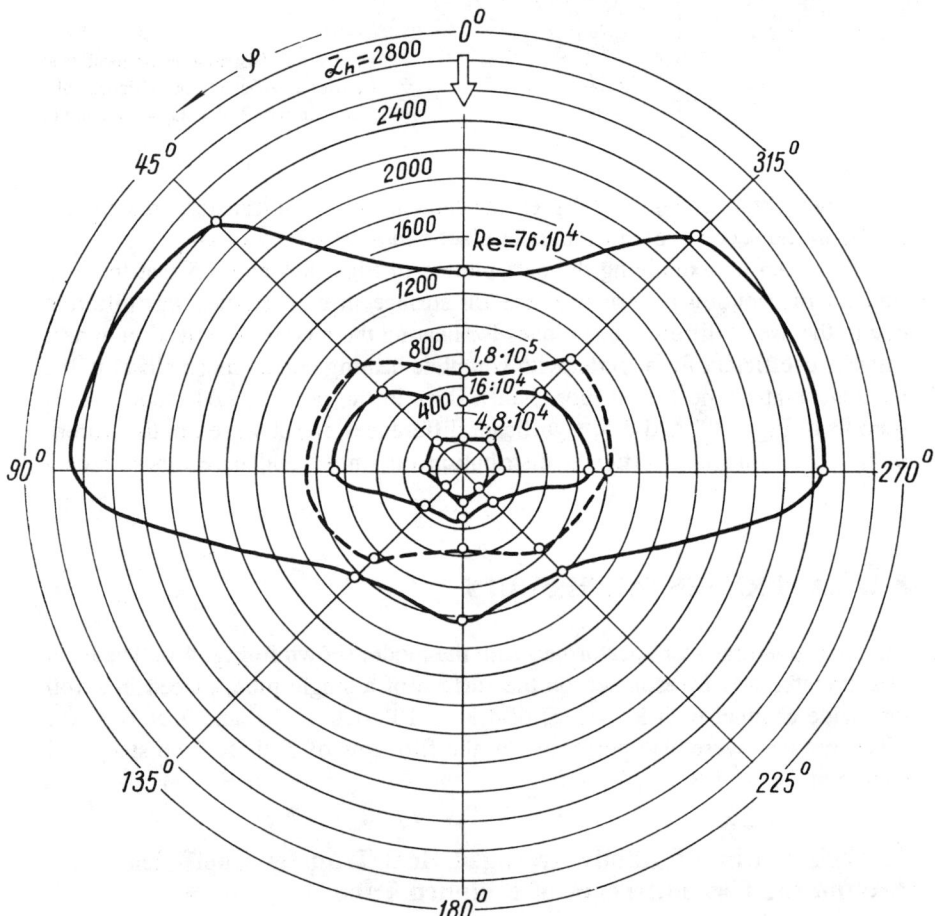

Figure 8.6 Distribution of a radially averaged heat transfer coefficient over the circumference of a finned tube ($s = 6$ mm) at different Re. The solid curves are for a single finned tube, the dashed curves are for a finned tube within a staggered bundle; *1)* Re $= 4.8 \times 10^4$; *2)* 1.6×10^5 *3)* 1.8×10^5; *4)* 7.6×10^5.

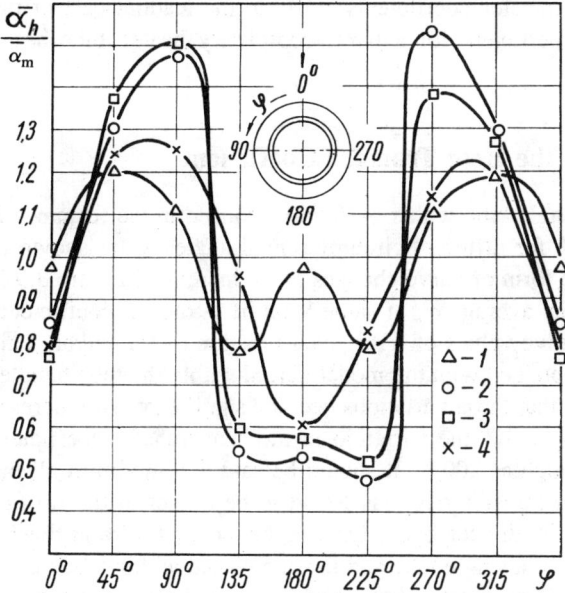

Figure 8.7 Variation of the relative heat transfer coefficient over the circumference of a finned tube ($s = 6$ mm) at different Re. *1)* Re $= 4.8 \times 10^4$, single tube; *2)* Re $= 1.6 \times 10^5$, single tube; *3)* Re $= 7.6 \times 10^5$, single tube; *4)* Re $= 1.8 \times 10^5$, in a bundle.

age. It will be seen from Fig. 8.6 that the coefficients are more uniformly distributed over the circumference in the bundle than they are for the single tube. In spite of the fact that the tube had spiral fins, there is no significant difference in the heat transfer coefficient between the two sides of the tube.

The same results, reduced in the form of relative heat transfer coefficients $\bar{\alpha}_h/\bar{\alpha}_m$ (where $\bar{\alpha}_m$ is the coefficient of heat transfer averaged over the height and circumference of the fin), are plotted in Fig. 8.7 in Cartesian coordinates as a function of angle φ. As is seen, the relative heat transfer coefficient increases with φ starting with the leading point, due to the increase in velocity adjacent to the tube surface in the flow direction, and attains a maximum near the equator (φ from 70 to 90°). Further round the tube it drops off steeply due to flow separation. At low Re, for the trailing part of a single tube, the relative coefficient of heat transfer increases somewhat. This effect is insignificant at higher Re. This effect was not detected in interior rows of a bundle, and in this case the trailing part is that with lowest rate of heat transfer. Note that high temperature fluctuations, up to 9.7K, were observed in this zone of the tube circumference.

It is seen from Fig. 8.7 that, at lower Re, the heat transfer coefficient is distributed more uniformly over the circumference than in cases with high Re. It is seen by comparing these data with similar results obtained for a bare tube with complete heating (see Fig. 8.1), that the greatest difference in the nature of distribution of local heat transfer coefficients exists in the leading part of the tube. The heat transfer coefficient for a bare, completely heated tube has a maximum at $\varphi = 0°$.

2. Radial Variation in the Heat Transfer Coefficient

Experimental data for a single tube with $s = 6$ mm, reduced in the form $\alpha_i/\bar{\alpha}h$ and plotted vs fin height h for different circumferential angles φ are shown in Fig. 8.8. It is seen that the form of curve changes as Re increases. Figure 8.9 is a plot of the circumferential average $\bar{\alpha}_\varphi$ of same local heat transfer coefficients plotted in the form of relative values of $\bar{\alpha}_\varphi/\bar{\alpha}_m$ as a function of fin height. This figure also contains data obtained with the model placed within the tube bundle. It follows from the graphs that these variations become significant with increasing Re, and at Re $= 7.6 \times 10^5$ the variation in the heat transfer coefficient along the fin height is as high as 200%. It should be said that in almost all the cases under study the heat transfer rate was found to be higher at the fin tip. This can be attributed to the thicker hydrodynamic boundary layers at the fin base. Comparison of data for a single tube and for one placed within a bundle at almost the same Re shows that the latter has a higher rate of heat transfer in the upper part of the fin, which is due to the turbulizing effect of the upstream rows.

3. Variation in Heat Transfer Coefficient as a Function of Fin Pitch

Graphical reduction of experimental data in the form $\alpha_i/\bar{\alpha}_h = f(h)$ at different fin pitches (Fig. 8.10) shows that the nature of distribution of local heat transfer coefficients over the fin height is different at different angles of rotation of the model, but depends little on fin pitch. Whereas at $\varphi = 0$ and $180°$, the maximum values of the relative heat transfer coefficients are attained at the fin tip, at $\varphi = 90°$ they are attained approximately in the middle of the fin height.

Upon selecting relative values $\bar{\alpha}_\varphi/\bar{\alpha}_m$ at different Re as a function of fin pitch, and plotting them, we obtain a curve of the variation in local heat transfer coefficients vs the fin pitch. Such a curve is shown in Fig. 8.11. It is seen that the heat transfer coefficient increases as the pitch is increased to 6 mm (mean distance between fins $u = 4.5$ mm). A further increase in pitch does not result in a perceptible increase in the heat transfer coefficient. The effect of the pitch on heat transfer decreases with increasing Re. Whereas, for example, at Re $= 4 \times 8 \cdot 10^4$ increasing the fin pitch from 4 to 7 mm results in a 14% rise in the relative heat transfer coefficient, at Re $= 1.6 \times 10^5$ it increases by 13%, and at Re $= 7.6 \times 10^5$ by only 7%. This is entirely logical, since the thickness of boundary layers decreases with increasing Re, and the maximum values of heat transfer coefficient are attained with smaller pitches.

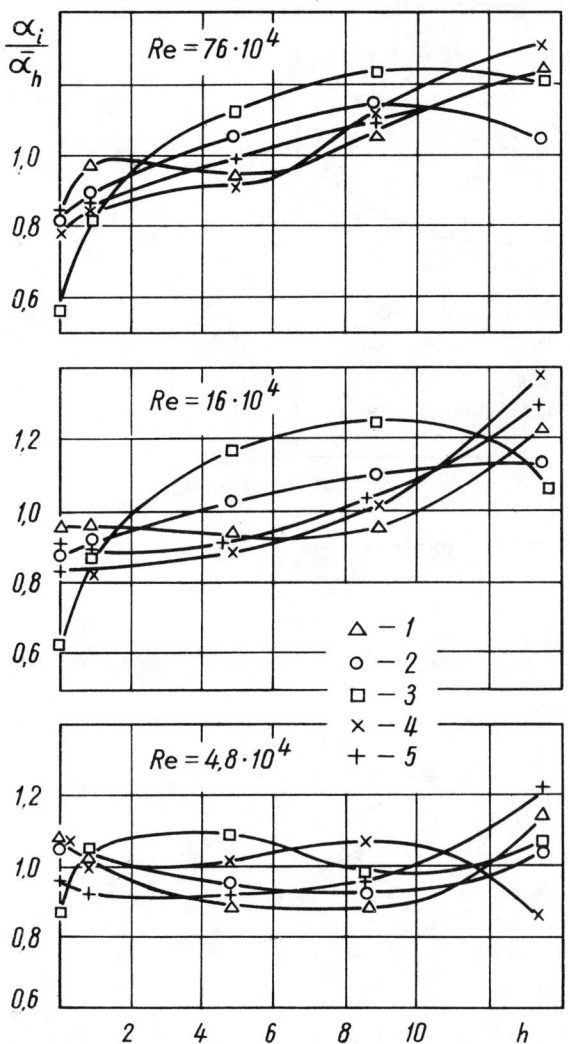

Figure 8.8 Variation in relative heat transfer coefficient over the height of a fin (s 6mm). *1)* $\varphi = 0$; *2)* 45°; *3)* 90°; *4)* 135°; *5)* 180°

148 HEAT TRANSFER OF FINNED TUBE BUNDLES IN CROSSFLOW

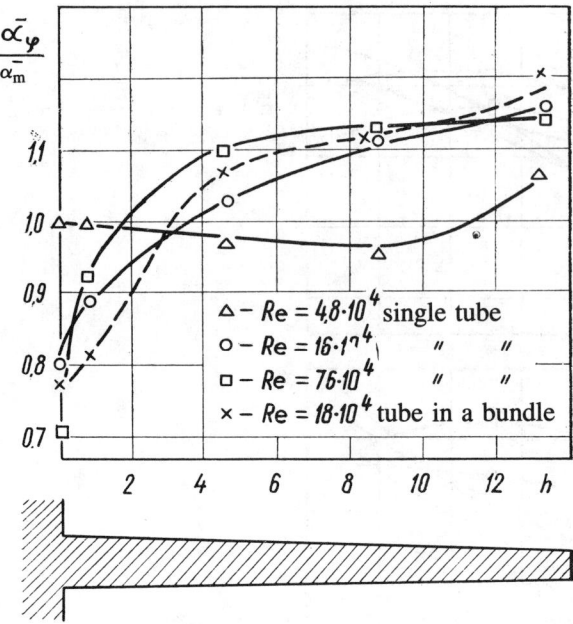

Figure 8.9 Variation in the mean relative heat transfer coefficients over the height of the fin at different Re ($s = \epsilon$ mm).

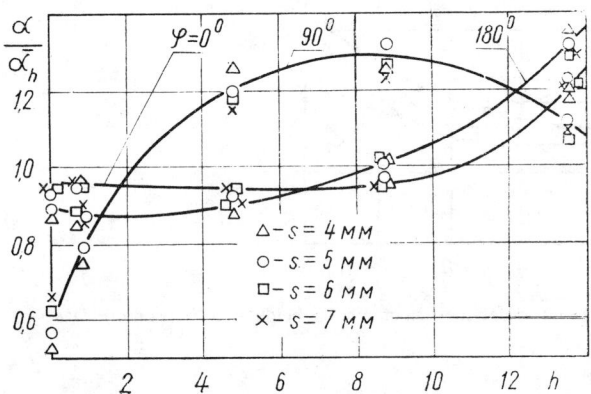

Figure 8.10 Effect of fin pitch on the distribution of the relative heat transfer coefficient over the fin height at different circumferential positions (Re = 1.6×10^5).

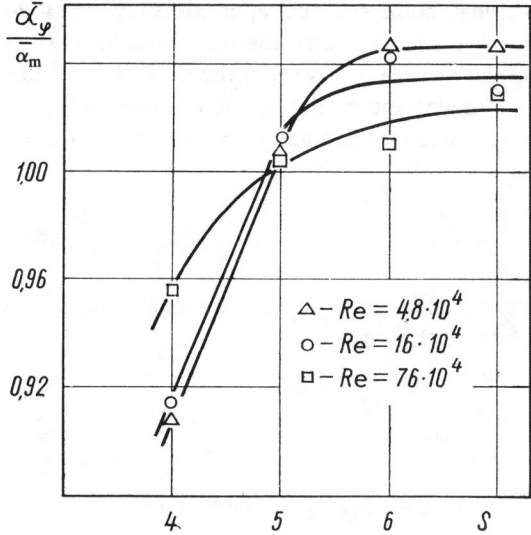

Figure 8.11 Effect of fin pitch on heat transfer.

These data are on the whole in satisfactory agreement with the data of Wegener [76]. However, it is difficult to agree with his suggestion that the optimum distance between fins be equal to one thickness of the boundary layer. It appears that this distance should be such as to prevent flow interaction between the fins, for which reason it should be at least as wide as two thicknesses of the boundary layer (see Fig. 8.12).

8.3. COMPARISON OF MODEL HEATING METHODS

In order to assess the local heating method use in the present investigation of local heat transfer, an approximate analytical method has been developed. This analysis is for a bare tube in crossflow and allows comparison between the complete and point heating methods, at least for this case.

Since we are interested only in the relative difference between the alternative heating methods, we shall restrict ourselves to the determination of the heat transfer distribution in a laminar boundary layer on a wedge, as suggested by Eckert [92]. The heat transfer coefficient at a point on a circular cylinder at a

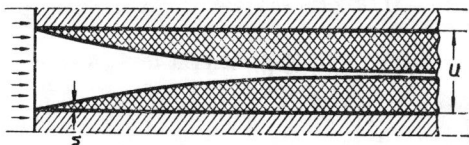

Figure 8.12 Boundary layers on surfaces of adjoining fins.

distance x from the leading critical point will be approximately equal to the heat transfer coefficient for a wedge at the same distance from its start, if the velocity immediately outside the boundary layer and the derivative of this velocity with respect to x on the cylinder and on the wedge are identical.

Since the velocity immediately outside the boundary layer on the wedge has a power-law distribution

$$w = w_0 x^m \qquad (8.2)$$

the assumed condition can be interpreted by a unique value of an exponent for a wedge as well as for a circular tube in the form

$$m = \frac{x}{w_0} \frac{dw_0}{dx} \qquad (8.3)$$

The velocity distribution at the leading part of a cylinder in crossflow is assumed according to Hiemenz [92]:

$$\frac{w_x}{w_0} = 3.6314 \left(\frac{x}{d}\right) - 2.1709 \left(\frac{x}{d}\right)^3 - 1.5144 \left(\frac{x}{d}\right)^5 \qquad (8.4)$$

Equation (8.4) is now used for calculating exponent m, which takes the form:

$$m = \frac{3.63 - 3 \cdot 2.17 x^2 - 5 \cdot 1.51 x^4}{3.63 - 2.17 x^2 - 1.51 x^4} \qquad (8.5)$$

The heat transfer from the leading part of the cylinder is now calculated using integral solutions of heat transfer from wedges with and without an initial heated region, presented by Tamonis [93]. Accordingly, the heat transfer on wedges at $\Pr = 1$ is determined from the expression

$$\left(\frac{\mathrm{Nu}_x}{\sqrt{\mathrm{Re}_x}}\right)^* = \frac{\mathrm{Nu}_x}{\sqrt{\mathrm{Re}_x}} \left[1 - \left(\frac{x_0}{x}\right)^l\right]^{-k} \qquad (8.6)$$

where

$$\frac{\mathrm{Nu}_x}{\sqrt{\mathrm{Re}_x}} = 0.379 (m+1)^{0.557} - 0.062 (m+1)^{-6.54},$$

$$l = 0.813 (m+0.93), \quad k = 0.376 - 0.039 (m+1)^{-6.18}$$

LOCAL HEAT TRANSFER COEFFICIENT FROM A FINNED TUBE IN A BUNDLE

The heat transfer with a nonheated initial region was calculated using the scheme shown in Fig. 8.13.

For convenience in calculation, quantities

$$\mathrm{Nu}_x/\sqrt{\mathrm{Re}_x}$$

(for the case of complete heating) and

$$(\mathrm{Nu}_x/\sqrt{\mathrm{Re}_x})^*$$

(for the case of a nonheated initial region) were multiplied by

$$\sqrt{w_0/x}$$

which made it possible to express the results in terms of nondimensional ratios, based on constant length x and free-stream velocity w_0.

The results of calculations represented Fig. 8.14 illustrates the difference in heat transfer on the leading side of a cylinder when heated by the two methods. It is seen that the heat transfer distributions are different. In the case of complete heating, the maximum heat transfer coefficient occurs at the leading part of the cylinder, whereas for point heating it is located at φ between 50 and 60°. As to the trailing side of the cylinder, one should not expect any differences here, since no stable boundary layer (velocity or thermal) forms following separation.

The above results of comparison can, to a first approximation, be extended to a finned tube.

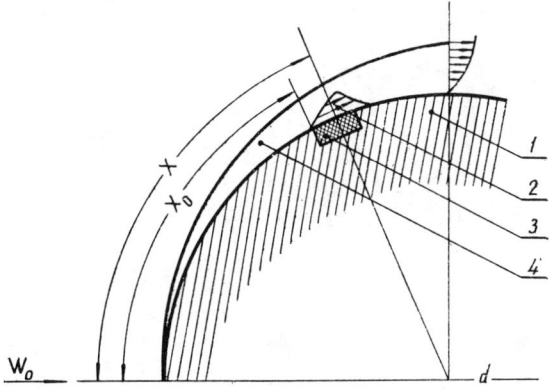

Figure 8.13 Schematic of circular surface with nonheated initial region. *1)* Tube; *2)* thermal boundary; *3)* heated element; *4)* velocity boundary layer.

152 HEAT TRANSFER OF FINNED TUBE BUNDLES IN CROSSFLOW

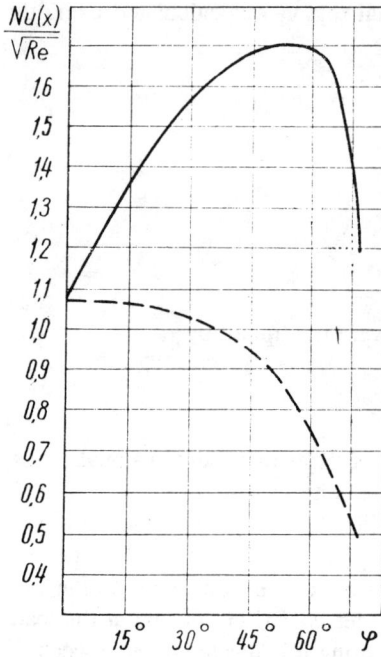

Figure 8.14 Distribution of local heat transfer coefficient over a circular surface, calculated analytically for a laminar boundary layer. The solid curve is for point heating and the dashed curve for complete heating.

CHAPTER
NINE

EFFECTIVENESS OF THE BUNDLES UNDER STUDY

The obtaining of nondimensional equations for a determination of the heat transfer and Euler number for bundles of finned tubes does not suffice for a definitive selection of the most effective finned surface for a given, specific application. In this chapter, we will discuss the various bases for comparison of various bundle and fin configurations in order to make such a selection.

9.1. PRELIMINARY REMARKS

The suggestion that the heat transfer coefficient and pressure drop for a heat-transmitting surface should both be compared in order to obtain an indicator of the overall thermal effectiveness of a heat exchanger is not new. The search for such an indicator resulted in obtaining the well-known Reynolds equation

$$\text{St} = \frac{\text{Nu}}{\text{Re Pr}} = \frac{\xi}{8} \qquad (9.1)$$

This analogy holds only at $\text{Pr} = 1$ and naturally has a very limited practical utility. This suggestion was pursued further by introducing the concept of the coefficient of analogy λ for a heat transmitting surface defined as

$$\lambda = \frac{8 \text{ St}}{\xi} \qquad (9.2)$$

(This coefficient was previously mentioned in Sec. 3.1.) A number of investigators suggested different expressions of λ for various forms of flow over bodies [88]. For example, the expression suggested by Hirschberg [42] was:

$$\lambda = \frac{1}{1+(\mathrm{Pr}-1)\,1.74\,\mathrm{Re}^{-1/8}} \quad (9.3)$$

At the same time, it was known that the selection of the optimum type of heat exchanger surface depends on the power expended in pumping the coolant, the heat exchanger dimensions and weight, and also on the fabrication costs. It has therefore become customary to select the most effect heat exchanger type on the basis of features of the thermal effectiveness of the surface and of dimensional and weight considerations. It should be noted that this is a rather difficult practical problem.

Comparison of the above parameters for convective heating surfaces is the subject matter of studies by Antuf'yev [10, 11, 61], Brauer [46, 17], Mitskevich [63] and Yudin [66]. The substance of all these studies is approximately the same; the comparison is performed at the same temperature difference between the coolant and the finned surface. In some cases all the parameters are referred to unit heat transfer surface [61, 66], and in others to unit heat exchanger volume [63].

In some cases, for example, in the paper by Brauer [17], the optimum fin height is found from experimental data by comparison with analogous quantities for a bare tube bundle. Though such representation of data is of interest, it reflects only the *thermal* effectiveness of the heat exchanger surfaces.

An extensively used method for comparing heating surfaces is that based on the energy factor E, introduced by Kirpichev [65]:

$$E = \frac{Q}{N} \quad (9.4)$$

where N is the pumping power. If the comparison is performed at $\Delta t = 1\,°\mathrm{C}$, and the pumping power consumption is referred to unit heat surface, this expression becomes

$$E_0 = \frac{\alpha_{\mathrm{red}}}{N_0} \quad (9.5)$$

Antuf'yev [10, 95] recommends the construction of graphs of $E_0 = f(N_0)$ or $\alpha = f(N_0)$ and the comparison of energy factors at constant N_0. The intercepts on the E_0 α_0 axes reflect the energy efficiency of a surface. Schmidt [96] suggests that the heat exchanger effectiveness be obtained from the expression:

$$E = \frac{8}{\xi} \frac{\text{Nu}}{\text{Re}} \frac{k}{\alpha} \qquad (9.6)$$

where k is the overall heat transfer coefficient. At $k/\alpha = 1$ the expression for the effectiveness approaches Reynolds equation (9.1). At $\text{Pr} = 1$ the heat exchanger effectiveness becomes equal to analogy factor λ.

In the papers by Mitskevich [63] and Yudin [66] the comparison is performed on the basis of the coefficient of effectiveness of the heated surface, which is defined as the ratio of the amounts of heat, transmitted from unit surface of the specimen and a reference surface at equal coolant pumping power per unit surface, and at equal temperature differences between the heating surface and the coolant. This coefficient incorporates the geometry, flow turbulization, thermal conductivity of the fin, and also the thermophysical properties of the coolant. The ratio of the volumes and weights in this case is obtained at constant Q, but at different pumping powers. The relationships suggested by the aforementioned investigators make it possible to compare the performance of various forms of surfaces, operating with different coolants and under different flow conditions.

Under our experimental conditions the question of comparison is simplified by the fact that the experiments were performed with the same finning type and with the same coolant.

9.2. COMPARISON OF BUNDLE EFFECTIVENESS ON THE BASIS OF THE SURFACE EXTENSION FACTOR

To determine which of the bundles is the most effective, we compared the thermal and hydraulic performance of finned and bare tube bundles. For this we used previously obtained data on heat transfer from nonfinned staggered bundles [20], the principal parameters of which are listed in Table 6.

Table 6 Geometric parameters of staggered bundles of bare tubes

Tube diameter d, mm	Bundle pitches s_1, мм	s_2, мм	a	b	Compactness factor Π, m²/m³	Weight factor B, kg/m²
32	70,4	41.6	2.2	1.3	54.3	11,28
23	57,0	29.4	2.48	1.28	43.1	14,2

156 HEAT TRANSFER OF FINNED TUBE BUNDLES IN CROSSFLOW

For this particular comparison technique, it is best to refer Nu_{red} to the surface area of the tube carrying the fins. This can be easily done by using the expression

$$\mathrm{Nu}' = \mathrm{Nu}_{red}\, \varphi \qquad (9.7)$$

If Nu_0 and Eu_0 correspond to bare-tube bundles, then ratios $\mathrm{Nu}'/\mathrm{Nu}_0$ and $\mathrm{Eu}/\mathrm{Eu}_0$ will show the relative rise in heat transfer and drag resulting from using finned tubes. The surface extension factor φ is an important parameter, which includes the fin geometry. Thus, if the aforementioned ratios are plotted as a function of φ, then the resultant curves (Fig. 9.1) will show the effect of the fin geometry on the heat transfer coefficient and pressure drop.

It is seen from Fig. 9.1 that the gain in heating transfer rate from bundles resulting from the use of taller fins is paralled by a significant rise in the pressure drop. At moderate φ the heat transfer rate from a finned bundle increases very steeply with h, whereas at higher φ the curve of $\mathrm{Nu}'/\mathrm{Nu}_0$ asymptotically approaches the horizontal, i.e., the reduction of the fin effectiveness starts being felt. This shows that a further increase in the fin height does not result in increasing the heat transfer rate from the bundle. The steeper rise in drag at high φ further emphasizes the disadvantage of continuing to increase the fin height.

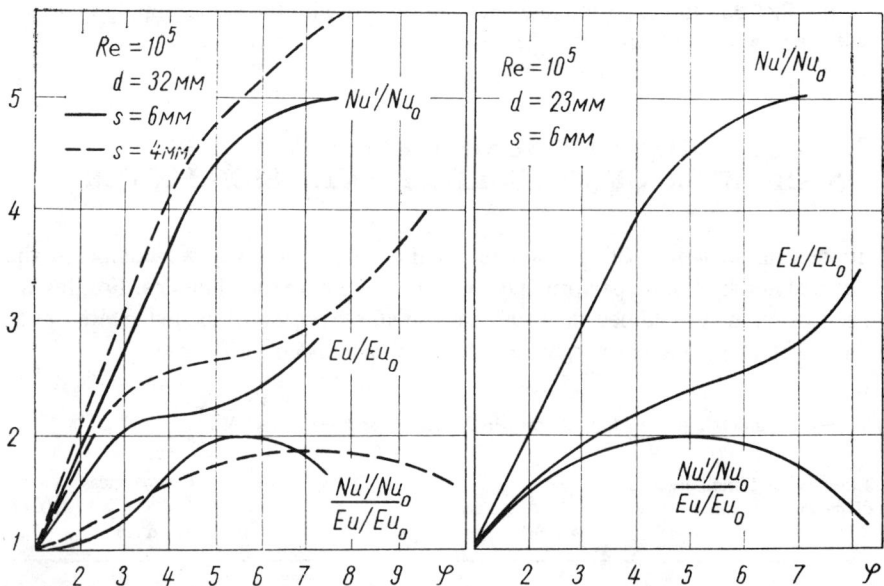

Figure 9.1 Variation in the relative indexes of heat transfer and draft of finned bundles as a function of the surface extension factor φ.

The ratio

$$\frac{Nu'/Nu_0}{Eu/Eu_0}$$

relates the gain in heat transfer rate from bundles, to the rise in pressure drop, i.e., it serves as an energy effectiveness factor for the bundle. The maximum value of this ratio indicates that the given bundle is optimal from the point of view of power effectiveness, i.e., its finning geometry is optimal. Figure 9.1 can be used for determining the value of φ corresponding to optimum fin pitch. Thus, for bundles of finned tubes with $d = 32$ mm the optimum value of fin height at $s = 6$ mm has a corresponding value of $\varphi = 5.3$. At $s = 4$ mm the optimum fin height corresponds to $\varphi = 7$ and for bundles with $d = 23$ and $s = 6$ mm to $\varphi = 5$.

Using curves of $\varphi = f(h)$ (which can be easily constructed for each bundle) we can determine the equivalent optimal fin height. Thus, for $d = 32$ mm it was found to be the following: $h = 10$ mm for $s = 6$ mm, and $h = 9.5$ mm for $s = 4$ mm. For $d = 23$ mm and $s = 6$ mm the optimum h was 8.7 mm.

The optimum finning geometry obtained in this manner does not incorporate the effect of weight and overall dimensions, and hence cannot be regarded as entirely valid. Nevertheless, these results deserve attention, since they provide a graphic idea on the effect of the thermal effectiveness of a fin on the optimum finning geometry. Note that it is extremely difficult to determine precisely the optimum parameters of finning for the general case. Since this can only be done by considering a large number of controlling factors, it is simplest to determine it separately for each heat exchanger on the basis of the process specifications.

9.3. COMPARISON OF THE THERMAL EFFECTIVENESS OF BUNDLES

The principal practical requirements in the design of a heat exchanger are the following: 1) the specified amount of heat should be transmitted by a heat exchanger with the minimum overall dimensions and weight; and 2) the blower power expended for moving the coolant should be at minimum. To satisfy these requirements in selecting an optimal bundle one must, in addition to determining the optimum fin height for each bundle described in the previous section, also know the thermal effectiveness for the various bundle configurations.

The thermal effectiveness is compared on the basis of the energy coefficient, calculated in the form of the ratio of the amount of heat released per unit time at 1 °C temperature difference to the power expended for overcoming the pressure drop, both the heat release and the pressure drop being referred and unit surface area (9.2). The energy coefficient can also be written in another form, more suitable for practical use:

$$E_0 = \frac{\alpha_{red}\varphi}{3.13\,\Delta p\,w_0 \cdot s_1/d \cdot \rho} \qquad (9.8)$$

where Δp is the pressure drop per tube row within the bundle, and w_0 is the free-stream velocity. The correction factor ρ in Eq. (9.8) makes allowance for the fact that the experiments being compared were performed at different air pressures.

Coefficient E_0 obtained at above-atmospheric pressures, is significantly higher than that obtained under standard conditions (p = 760 mm Hg, and t_f = 20 °C). This indicates that it is desirable to employ coolants at high pressure; this is illustrated by the results show in Figure 9.2 which is a plot of the energy coefficients for bundle 6, determined under standard-pressure conditions (solid line) above-atmospheric pressure conditions (triangles) the latter being calculated according to Eq. (9.8).

The complications resulting from the comparison of coefficients E_0 of different bundles obtained under different experimental conditions can be avoided by reduction to identical conditions by establishing the pressure dependence of E_0 by an approximate method. E_0 was found to be directly proportional to the absolute coolant pressure to a power of 0.8. This can be written in the form:

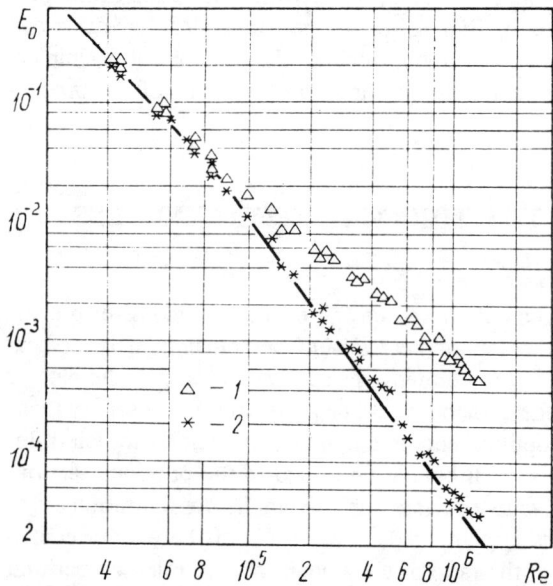

Figure 9.2 Plot of $E_0 = f(Re)$ of bundle 6. The solid line represents the energy coefficient for standard pressure coolant. 1) Points corresponding to above-atmospheric pressure conditions; 2) points reduced with Eq. (9.9).

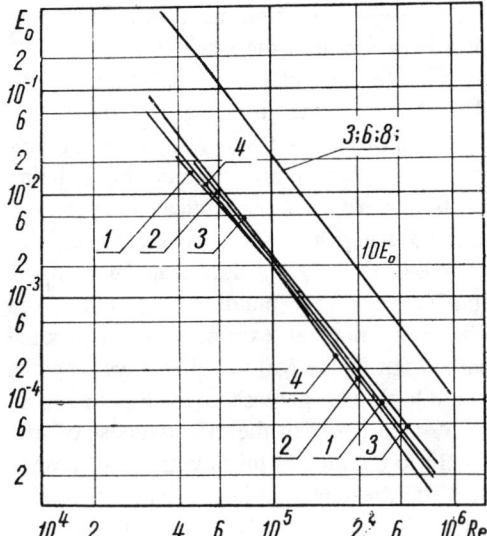

Figure 9.3 Comparison of thermal effectiveness of bundles as a function of Re.

$$E_{0g} = E_{0st} p^{0.8} \quad (9.9)$$

where p is the absolute pressure in bars, subscript g designates conditions at the latter pressure, and st pertains to standard coolant pressure.

The above-atmospheric pressure experimental points for bundle 6 were corrected for pressure using Eq. (9.9) and are also plotted (as asterisks) in Fig. 9.2; these fit satisfactorily on the line for standard conditions. It can be concluded from Fig. 9.2 that, from the viewpoint of thermal effectiveness of a bundle it is always desirable to increase the coolant pressure. A subsequent comparison of bundles was performed on the basis of energy coefficients, reduced to standard conditions using Eq. (9.9).

Figure 9.3 compares the thermal effectiveness (as a function of Re) for bundles of tubes with different fin heights (h varying) at constant s (bundles 1 through 4) and different fin pitches (s varying) at constant h (bundles 3, 6 and 8). It is seen from comparison of the first set of bundles, in the lower part of the graph, that at constant Re, bundle 3 exhibits the best performance, energywise. Bundle 4, with the highest fins, has a lower thermal effectiveness, particularly at high Re. This can be attributed to the fact that high fins, situated in a forced coolant flow, have low fin effectiveness values. At higher Re, bundles of tubes with low fins (bundle 1) have higher energy effectiveness values. Apparently, at high Re the effectiveness of low fins decreases less. Comparison of the second group of bundles (curves plotted in the upper part of Fig. 9.3) shows that, from the energy point of view, they are equivalent.

This comparison, however, does not adequately represent the actual thermal effectiveness of bundles, since comparison for the case of constant Re makes no allowance for the difference in specific pumping power, since the pressure drop for the different bundles is different at the same Re.

It is somewhat simpler to compare the thermal effectiveness of bundles by the technique due to Antuf'yev [10]. In this case the comparison is performed with graphs, expressing function $E_0 = f(N_0)$ or $\alpha_{red} = f(N_0)$. The energy parameters equalize at given N_0. This comparison also has its shortcoming of being performed at different Re. Nevertheless, it is of greater practical interest.

Henceforth the bundles are compared using the Antuf'yev technique. Figure 9.4 shows curves for bundles, plotted as $\alpha_{red} = f(N_0)$. It is seen, by comparing bundles at constant N_0 with given s but varying h (bundles 1 through 4, and also 9 through 11), that bundles with low fin tubes exhibit the best thermal performance. It can be noted here that at high N_0 (or Re) the difference between the performance of bundles with low and high fins increases, which points to an increase, with a rise in Re, in the difference between the effectiveness of low and high fins. However, this applies only to bundles of tubes with $d = 32$ mm. Comparison of bundles with varying s and the same fin height under the same conditions (Fig. 9.4) points to a small advantage of bundle 8, which has the largest s.

The same graph also presents similar data for bundles of non-finned tubes with $d = 32$ and 23 mm (for the specifications of these bundles see Table 6). It is seen that the thermal effectiveness of bare tube bundles is somewhat lower than that of bundles with low fins, but higher than that of bundles of tubes with high fins. Also, the thermal effectiveness of bare tube bundles (with $d = 32$ mm) increases somewhat more, with a rise in N_0 (or Re), than that of finned bundles. The only exception is bundle 1 with low fins, the fin effectiveness of which is close to 1.

It is seen by comparing the effectiveness of bundles 13 through 21 that the difference between them is rather small, with a slight advantage held by denser packed bundles. These bundles have a thermal effectiveness lower than that of bundles of bare tubes.

9.4. OVERALL DIMENSIONS AND WEIGHT OF BUNDLES

Using the data of Fig. 9.4 at a given N_0 we can determine the degree of thermal effectiveness of the bundles being compared. The latter is the ratio of the heat transfer coefficients:

$$\psi_i = \frac{\alpha_{red}}{\alpha_{red.r}}$$

where subscript r designates the reference bundles, whereas i refers any of the bundles being compared.

It is possible to determine the overall dimensions and weight of a bundle with a given heat transmission capacity. Assuming constant Q and $\Delta t = 1\,°C$, we can write:

EFFECTIVENESS OF THE BUNDLES UNDER STUDY 161

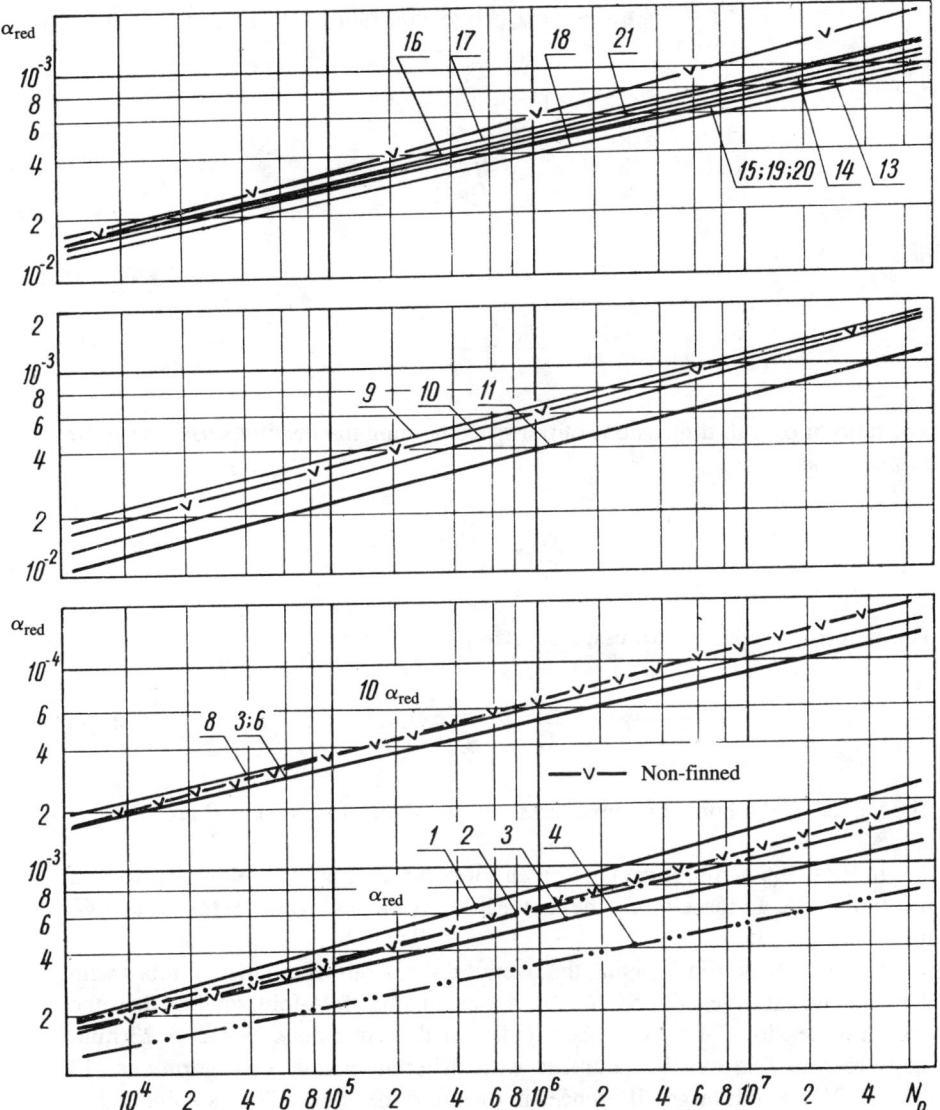

Figure 9.4 Comparison of the thermal effectiveness of bundles in the form $\alpha_{red} = F(N_0)$.

$$F_i \alpha_{\text{red}\,i} = F_r \alpha_{\text{red.r}} = Q = \text{constant}$$

whence

$$\frac{F_i}{F_r} = \frac{\alpha_{\text{red.r}}}{\alpha_{\text{red}}}$$

or

$$\frac{F_i}{F_r} = \frac{1}{\psi_i} \qquad (9.11)$$

The ratio of overall dimensions of the space holding the heating surfaces can be expressed as

$$\frac{v_i}{v_r} = \frac{\Pi_r}{\Pi_i \psi_i} \qquad (9.12)$$

The ratio of weights is expressed similarly:

$$\frac{G_i}{G_r} = \frac{B_i}{B_r \psi_i} \frac{1}{} \qquad (9.13)$$

The values of the bundle compactness and weight factors Π and B are listed in Table 7.

In Figs. 9.5 and 9.6, data are given for compactness and weight factors as a function of N_0. In these plots, the factor referred to analogous factors v_0 and G_0 for a bare tube bundle.

It is seen from Fig. 9.5 that the overall dimensions of bundles of tubes with $d = 32$ mm at given Q and N_0, are approximately five-fold smaller than the overall dimensions of a bare tube bundle for the same duty. For $d = 23$ mm, they are four-fold smaller. Bundles with different bundle configurations (13 through 21) exhibit large differences in overall dimensions. This is quite understandable, since the overall dimensions of a bundle are proportional to its packing density.

For given values of Q and N_0, it was found by comparing bundles on the basis of their weight (Fig. 9.6) that, over the range of fin geometry and Re covered here, the weight of a finned tube is from 0.5 to 1.2 of the weight of a bare tube bundle, with the lowest values being those of bundles of 23 mm diameter tubes. The lowest-weight tubes are those of tubes with shorter fins and large fin pitches. The weights of bundles with different configurations (13 through 21) are rather close.

Table 7 Supplementary data on the bundles under study

Bundle number	Least flowpassage area in a row, $f \cdot 10^2$, m²	Surface extension factor	Compactness factor	Weight factor kg/m²
1	3,54	3,09	106	7.76
2	3,39	4,42	151	7.04
3	3,39	6.25	198	7,01
4	3,39	10.6	301.5	6.12
5	3,54	3,29	112.5	7.50
6	3,39	4.78	164	6.80
7	3,71	7.35	209	6,30
8	3,39	3,86	133	7.20
9	3,54	3.47	123	8.24
10	3,39	5,15	180	7.46
11	3,97	8,2	220	6,66
12	3,33	3,18	111	8.80
13	1,75	5.23	128.2	7.90
14	2,59	5.23	99.2	7.90
15	2,59	5,23	89.2	7.90
16	2,09	5,23	123,3	7,90
17	2,09	5,23	159.2	7,90
18	2,65	5,23	81,3	7.90
19	2,31	5,23	95,1	7,90
20	2,00	5,23	183,2	7,90
21	2,67	5,23	118.7	7.90

There is no point in giving preference to one of the aforementioned factors (i.e., thermal, dimensional or weight) as a basis for heat exchanger selection, without introducing cost considerations.

9.5. EFFECT OF FINNING GEOMETRY ON THE THERMAL PERFORMANCE, OVERALL SIZE, AND WEIGHT OF BUNDLES

The thermal effectiveness of bundles can be expressed as $E_0 = f(N_0)$, and comparisons made at the given N_0. Thus, the degree of thermal effectiveness of bundles is expressed as the ratio of the corresponding energy coefficients:

164 HEAT TRANSFER OF FINNED TUBE BUNDLES IN CROSSFLOW

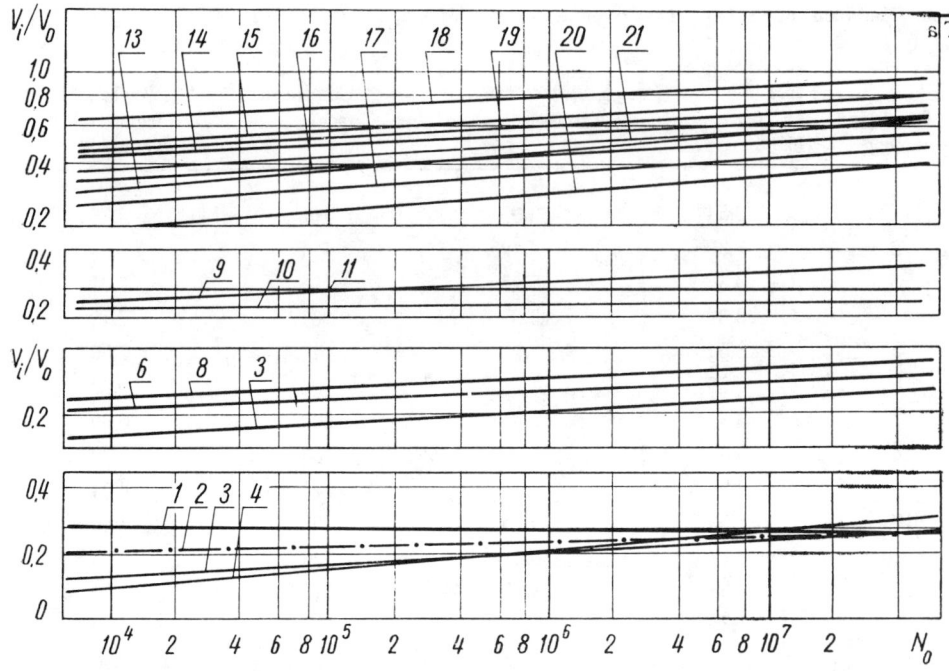

Figure 9.5 Comparison of bundles on volume basis.

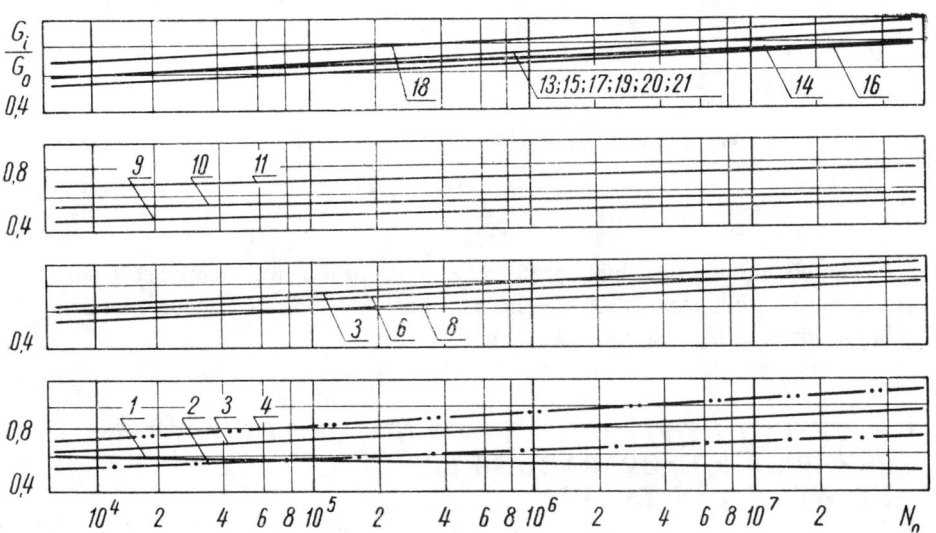

Figure 9.6 Comparison of bundles on weight basis.

$$\psi'_i = \frac{E_i}{E_r} \qquad (9.14)$$

The overall dimensions of bundles can be characterized by the ratio

$$\frac{N_0}{\Pi_i \, \psi_i}$$

which designates the overall dimensions of the bundle at a given heat transfer rate and coolant pumping power. The weight of bundles for analogous conditions can be expressed by ratio

$$\frac{N_0 \, B_i}{\psi'_i}$$

Then, referring the thermal performance, overall size and weight factors successively to the height and pitch of bundles under study, an indication can be obtained of the effect of the finning parameters on these quantities.

1. Effect of Fin Height

Figure 9.7 shows the thermal, size and weight factors of the bundles under study as a function of fin height for three values of Re.

These graphs reflect the dependence of the above factors on the range of Re over which the heat exchanger is to operate. Thus, fins with $h = 9$ mm are seen to be advantageous at $\text{Re} \leq 5 \times 10^4$ for both tube diameters (23 and 32 mm). A heat exchanger with such a fin has the best thermal performance, minimum overall dimensions and almost minimum weight.

At higher Re the optimum value of h is shifted somewhat toward lower values. At $\text{Re} = 5 \times 10^5$ it is best to use fins with $h = 9$ for $d = 32$ mm, and $h = 7$ mm for $d = 23$ mm.

2. Effect of Fin Pitch

Figure 9.8 shows the thermal, size and weight factors for bundles 3, 6 and 8 (at $h = 9$ mm = const) as a function of the fin pitch at three values of Re. It is seen that the best pitch for use at 5×10^4 is 6 mm. The optimum values of the pitch also shift to the lower values with increasing Re. At $\text{Re} = 5 \times 10^5$ it is best to use $s = 5$ mm.

It is seen that large pitches, for example, $s = 8$ mm, are less suitable, since they exhibit poorer thermal performance and involve an increase in the overall bundle dimensions at all the values of Re.

It must be remembered, in analyzing the question of the optimum pitch, that in the final analysis, it is the distance between fins u, rather than the fin

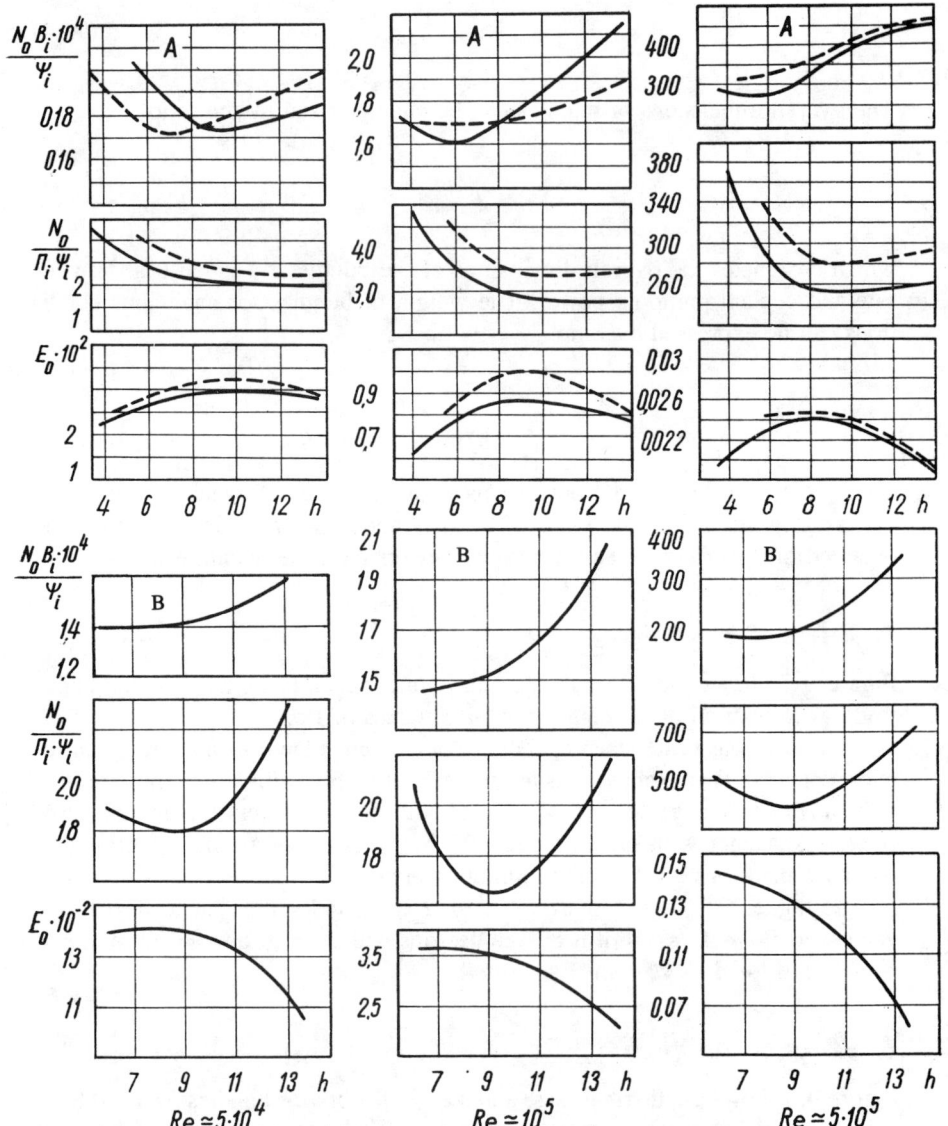

Figure 9.7 Thermal performance, size and weight factors as a function of fin height at different Re. *A)* Bundles of tubes with $d = 32$ mm, solid curve $s = 4$ mm; dashed curve $s = 6$ mm; *B)* bundles of tubes with $d = 23$ mm, $s = 6$ mm.

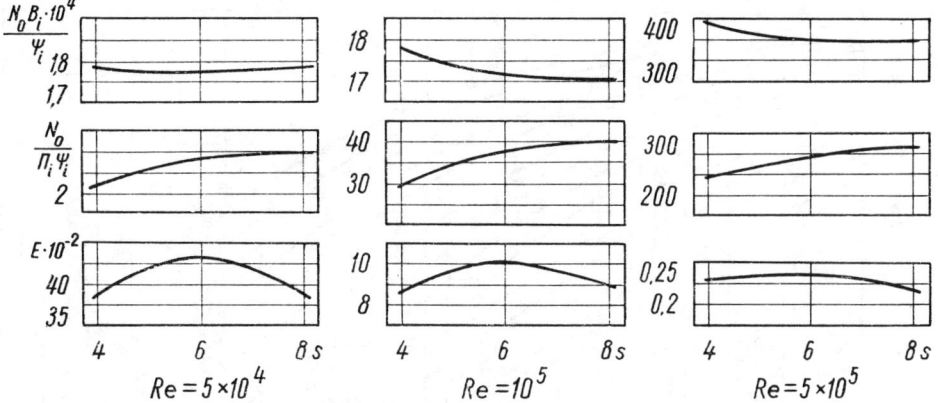

Figure 9.8 Thermal performance size and weight factors as a function of fin pitch s (at fin height $h = 9$ mm) and different Re.

pitch, which is the controlling factor, since, for example, the distance between fins is additionally controlled by the fin thickness. Hence the optimum distance between fins at Re $= 5 \times 10^4$ is actually $u = 4.5$ mm. At higher Re ($\sim 5 \times 10^5$) $u < 4.5$ mm.

3. Effect of Fin Thickness

All the bundles under study employed fins of the same thickness, and hence the effect of thickness on thermal performance and other factors was not determined experimentally. However, the effect of fin thickness can be clarified with sufficient accuracy by analytical methods.

It is known that the convective heat transfer coefficient is virtually unaffected by changes in the fin thickness. Only the fin effectiveness changes, and it, in its turn, has a significant effect on α_{red}. It should also be added that while the fin thickness has almost no effect on the bundle dimensions, it affects its weight rather significantly.

Hence, in order to determine the effect of fin thickness on heat transfer from a bundle, the fin effectiveness should be calculated over a wide range of thickness values and as a function of α (and thus of Re). This will allow one to judge, on the basis of these calculations, the advisability of using a fin with a given thickness. The effectiveness for fins of different thickness was determined from Fig. 2.7. The fin thickness was represented by the factor

$$\beta h = \sqrt{\frac{2\alpha}{\lambda \delta}}\, h$$

Figure 9.9 is a plot of fin effectiveness E as a function of Re for trapezoidal

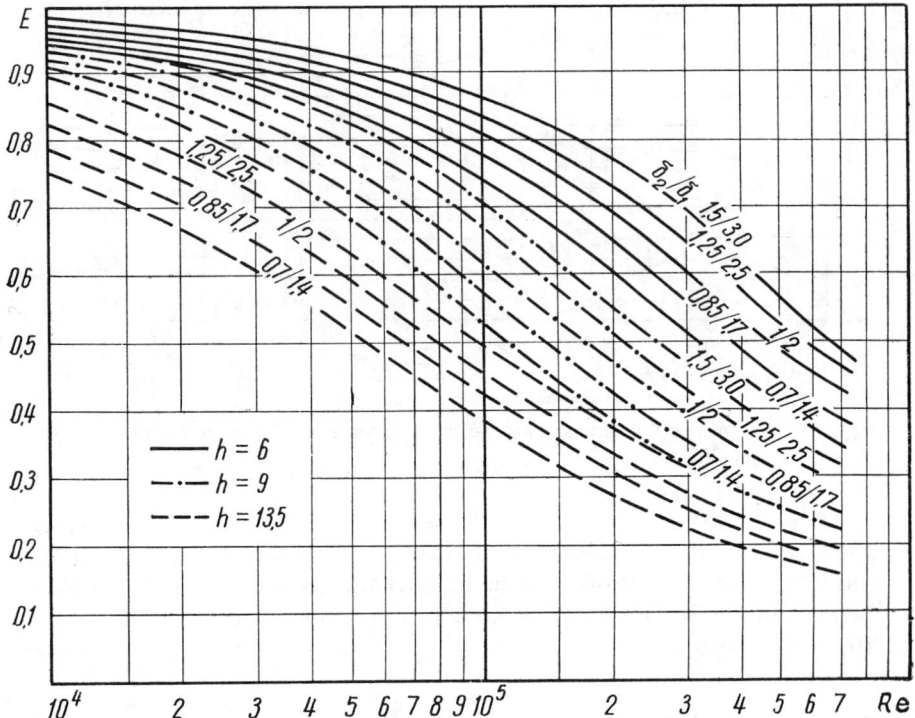

Figure 9.9 Fin effectiveness for different fin thickness as a function of Re. (δ_2, and δ_1 are in mm.)

spiral fins of various thicknesses and fin height. Here, the ratio of δ_2 (the tip thickness) to δ_1, (the base thickness) is always 0.5 (as in the present experiments); the results for the presents data (δ_2 = 1 mm, δ_1 = 2 mm) are included in the figure. Using this graph one can easily find ratios of fin effectiveness E/E_{exp} relating the value of effectiveness for a given thickness to that determined in the present experiments. The correction coefficient ξ, accounting for the trapezoidal shape of the fin, was not used in compiling Fig. 9.9 and neither was the nonuniformity in the distribution of heat transfer over the fin taken into account. This was because the object was to calculate E/E_{exp}^* and the factors cancel out.

It is seen by examining Fig. 9.9 that the gain due to increasing fin thickness (at h = 6 and 9 mm) at high Re (from 10^5 to 10^6) is much greater than at lower Re; however, this difference becomes smaller with increasing h. A curve was then constructed of $\alpha_{\text{red}}/\alpha_{\text{red exp}} = f(E/E_{\text{exp}}$, taking account of variations in the fin effectiveness, which is different for each value of h (Fig. 9.10). From this graph we determined, for a given E/E_{exp}, the value of $\alpha_{\text{red}}/\alpha_{\text{red exp}}$ corresponding to the analytic and experimental reduced heat transfer coefficients for fins with different h. The value of α_{red} can then be easily obtained from this ratio. To

EFFECTIVENESS OF THE BUNDLES UNDER STUDY 169

simplify the calculations, Fig. 9.10 was constructed from the simplified expression of $\alpha_{red} = f(\alpha)$ (Eq. 6.10), from which it follows that:

$$\frac{\alpha_{red}}{\alpha_{red.r}} = \frac{E_t + F_{fin}E\xi}{F_t + F_{fin}E_{exp}\xi_{exp}} \qquad (9.15)$$

The coefficient ξ, which corrects for the trapezoidal shape of the fin, was obtained from Fig. 3.3.

Figure 9.11 shows how the fin thickness and the (related) weight of one square meter of heating surface changes as a function of the convective heat transfer coefficient and fin effectiveness. This graph was constructed from data for bundle 10, which exhibits the optimal performance.

It is seen from the graph that at low α (or Re) the optimum fin thickness (beyond which little increase in effectiveness occurs) is $\delta_2/\delta_1 = 1/2$, which corresponds to a mean thickness of $\delta_m = 1.5$ mm. At higher α (i.e., Re) the effectiveness continues and increase with increasing thickness and the optimum thickness may be higher. However, must be added that at high Re or α the advisability of using thicker fins can be determined only on the basis of cost considerations. The calculations described in this chapter were performed for fins on a steel finned tube with a metal thermal conductivity of 46.5 w/mxk.

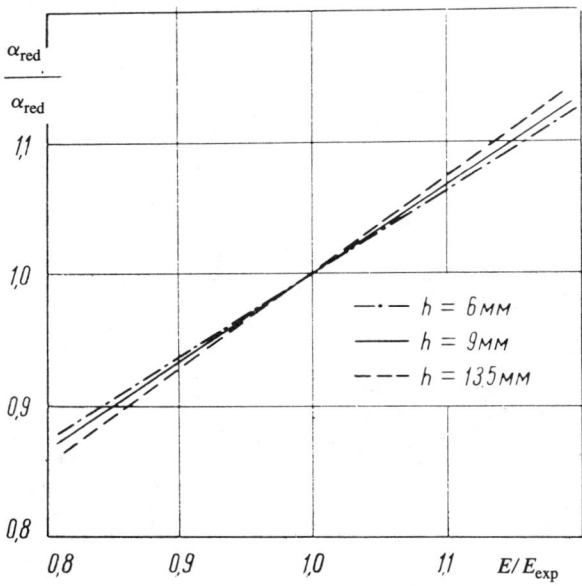

Figure 9.10 Graph for recalculating α_{red} as a function of fin effectiveness.

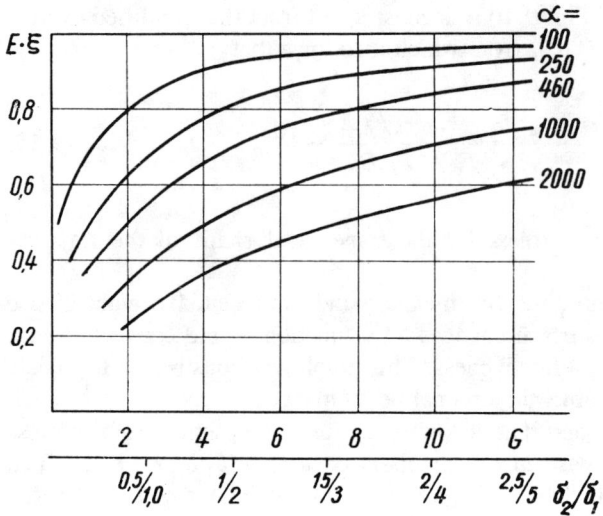

Figure 9.11 Effectiveness of trapezoidal fin as a function of fin thickness and weight of the finned tube with α as a parameter for bundle 10 ($d \times s \times h = 23 \times 6 \times 9$).

CHAPTER
TEN

COMPARISON OF DATA AND PRACTICAL RECOMMENDATIONS

This concluding chapter is concerned with a comparison of the present experimental data with those obtained by other investigators and with the presentation of recommendations for calculating the heat transfer coefficient and pressure drop for finned tube bundles.

10.1 COMPARISON OF DATA ON HEAT TRANSFER FROM FINNED TUBE BUNDLES

The relationships for which a wide range of Re, obtained by us for correlation of experimental data, make it possible to suggest practical recommendations on calculating heat transfer and drag of finned bundles.

Initially, we compared our correlations in the form Nu = f(Re) with experimental data of others, this data being reduced using the same technique.

The first comparisons were performed for the staggered bundle data of Yudin and Tokhtarova [77] for tubes with spiral fins with the following geometries: $d \times s \times h \times \delta$ = 23 × 5 × 5 × 1 mm, $a \times b$ = 3 × 1.2; $d \times s \times h \times \delta$ 23 × 2.5 × 5 × 0.6 mm, $a \times b$ = 2.82 × 1.5 and $d \times s \times h \times \delta$ = 23 × 5 × 10 × 1 mm, $a \times b$ = 3 × 1.2. The data are presented in Fig. 10.1 (points 1 through 3); for Re > 2 × 10^5, there is a more rapid increase in Nu with Re, which points to the predominance of turbulent flow over the bundle. Also shown in this figure is the curve obtained from the present work. The Yudin and Tokhtarova data agrees with our curve for mixed flow (Re < 2 ×

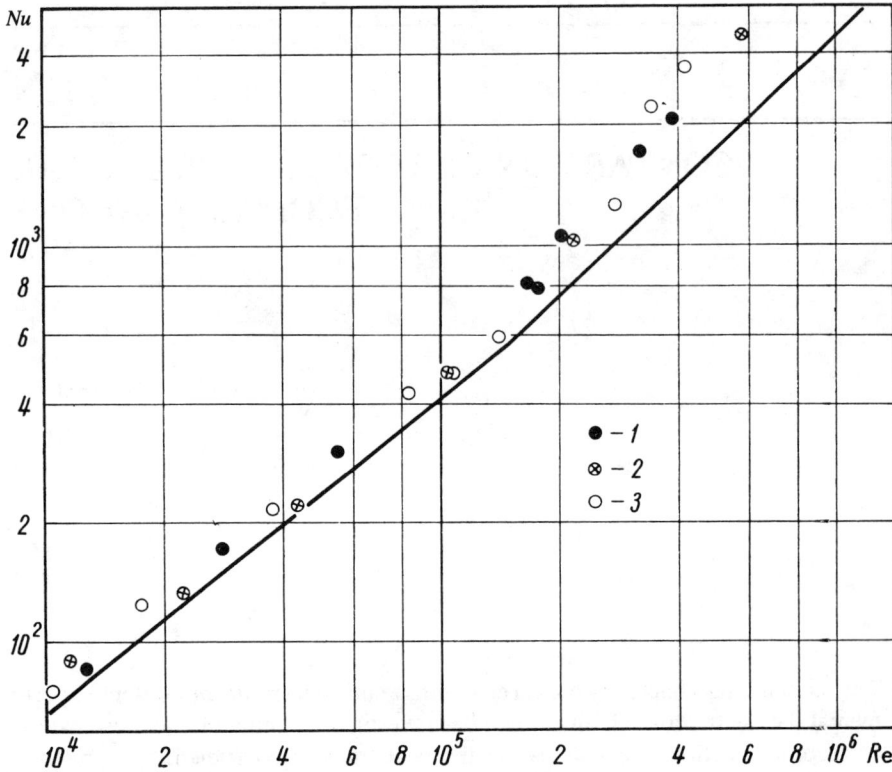

Figure 10.1 Comparison of data Yudin with Tokhtarova [77] with our correlations on heat transfer from staggered finned bundles. The solid curve represents our equations (7.4) and (7.5). *1)* $d \times s \times h \times \delta = 23 \times 5 \times 5 \times 1$ mm, $a \times b = 3 \times 1.2$; *2)* $23 \times 5 \times 10 \times 1$ mm, 2.82×1.5; *3)* $23 \times 5 \times 10 \times 1$ mm, 3×1.2.

10^5) but there is an increasing discrepancy with increasing Re for Re $> 2 \times 10^5$, this discrepancy being as high as 60% as the highest Re.

Our correlations (Eqs. (7.4) and (7.5)) were next compared with equations obtained by Schmidt [37] and Yudin, et. al. [38, 77]. Schmidt suggested the following nondimensional equation for staggered bundles

$$\text{Nu} = 0.45 \, \text{Re}^{0.625} \left(\frac{F}{F_t} \right)^{-0.375} \text{Pr}^{1/3} \qquad (10.1)$$

and the following for in-line bundles

$$\text{Nu} = 0.30 \, \text{Re}^{0.625} \left(\frac{F}{F_t} \right)^{-0.375} \text{Pr}^{1/3} \qquad (10.2)$$

where the reference parameters are the fin root diameter (diameter of the base tube) and the flow velocity in the most constricted flowpassage; however, the convective heat transfer coefficient was referred to the total surface of the tube.

The nondimensional equations suggested by Yudin, et. al. [38] are for staggered bundles

$$\mathrm{Nu} = 0.23\, k_z\, \varphi^{0.2} \left(\frac{d}{s}\right)^{-0.54} \left(\frac{h}{s}\right)^{-0.14} \mathrm{Re}^{0.65} \tag{10.3}$$

and for in-line bundles

$$\mathrm{Nu} = 0.105\, k_z\, k_s \left(\frac{d}{s}\right)^{-0.54} \left(\frac{h}{s}\right)^{-0.14} \mathrm{Re}^{0.72} \tag{10.4}$$

In the above expressions the convective heat transfer coefficient referred to the total surface of the finned tube, and the physical properties were determined at the mean flow temperature. The fin pitch was used as the reference dimension in Nu and Re. The flow velocity is determined in the narrowest cross section of the bundles; k_z is a correction factor for the number of longitudinal rows, and obtained from graphs (see Figs. 3.12 and 3.13) and k_s is a correction factor for the arrangement of tubes within the bundle. At $b < 2$ the values of k_s are determined from the nomogram in [38], whereas at $b > 2$, $k_s = 1$.

The comparison with the Schmidt Yudin et. al. equations were performed for four bundles with different finning parameters and different configurations (1) $d \times s \times h = 32 \times 4 \times 13.5$ mm, $a \times b = 2.38 \times 1.46$; (2) $d \times s \times h = 32 \times 8 \times 9$ mm, $a \times b = 2.17 \times 1.27$; (3) $d \times s \times h = 23 \times 4 \times 4$ mm; $a \times b = 2.67 \times 1.46$ and (4) $d \times s \times h = 23 \times 6.5 \times 10$ mm, $a \times b = 4.11 \times 2.14$. Equations (10.1)–(10.4) were used to determine for each bundle, Nu = f(Re) with Nu and Re being defined as in the present work, appropriate corrections being made for the differences in the reference dimensions employed by the different investigators.

Figure 10.2 illustrates the results. It is seen that the curves obtained by Schmidt and Yudin are rather close. At Re from 10^4 to 2×10^4 they are in satisfactory agreement with Equation (7.4). Note that, in this range of Re, exponent m changes from 0.625/0.65 to 0.8; this apparently due to the increased level of turbulence of the mixed flow in the boundary layer. However, due to the splitting of the flow by the fins, this boundary layer turbulence does not result in large increases in turbulence in the mainstream flow. The turbulence of this flow does, however, increase with Re, but only at Re $\simeq 2 \times 10^5$, as in smooth tube bundles, does transition occur to develop turbulent flow, accompanied by a rise of m 0.95 and, naturally, by a further enhancement of heat transfer.

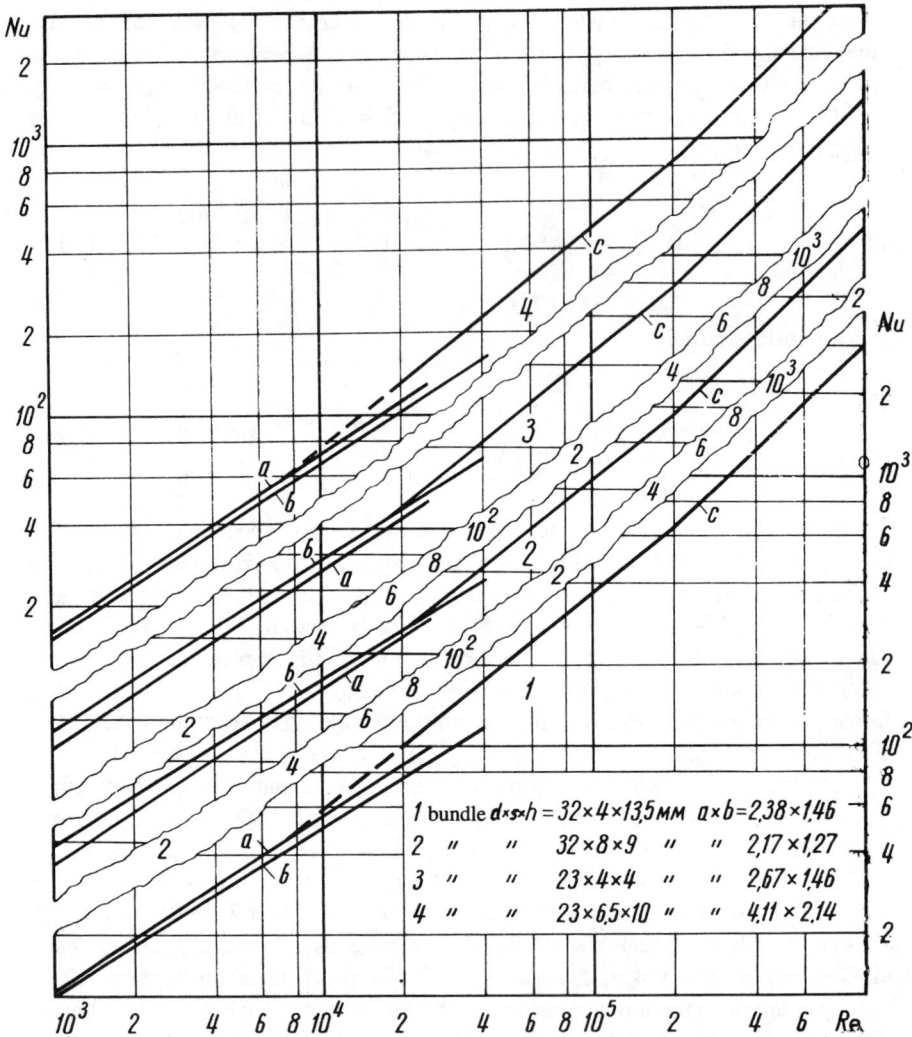

Figure 10.2 Comparison of our correlations on heat transfer from staggered finned bundles with those of other investigators. *a)* Results of Yudin [38] Eq. (10.4); *b)* results of Schmidt [37], Eq. (10.1); *c)* our equations (7.4) and (7.5).

10.2 COMPARISON OF DATA FOR THE EULER NUMBER FOR FINNED TUBES

Our correlations (Eqs. (5.4) and (5.5)) for the Euler number were compared in a manner similar to that used for heat transfer, with data obtained by Yudin and Tokhtarova [60, 86], Brauer [17] and Mirkovič [36]. The experimental data taken from these studies were recalculated in the form used in the present correlations. The comparison was performed for bundles of finned tubes with geometries as close as possible to our bundles. We compared data on staggered bundles with the following parameters: $d \times s \times h \times \delta = 23 \times 5 \times 5 \times 1$ mm and $a \times b = 3 \times 1.2, 23 \times 2.5 \times 5 \times 0.6$ mm, 2.82×1.5 from [60], then $32 \times 6 \times 9 \times 1.3$ mm, 2×1.2 and $32 \times 6 \times 9 \times 1.3, 3 \times 1.5$ from [86] (all the four bundles selected had spiral fins). In addition, we compared two bundles with circumferential fins, namely that studied by Brauer [17] with $d \times s \times h \times \delta = 28 \times 5.5 \times 5.5 \times 1.5, a \times b = 1.8 \times 1.8$ according to and that used by Mirkovic [36] with $d \times s \times h \times \delta$ $25.4 \times 4.2 \times 9.5 \times 1.27, a \times b = 4 \times 2.3$.

These data are compared in Fig. 10.3, from which it is seen that the agreement between the present correlations and the data is satisfactory, the error does not exceed the limits of accuracy of the correlations as determined in the reduction of our own data. No comparison was made for Re $> 5 \times 10^5$, since no data were available for this range.

Given the fact that the bundle configurations were not of the same type, and also that there were differences in the measuring techniques, in inlet conditions etc., it can be concluded from the above comparisons that our correlations are well validated for the calculation of the Euler number (and hence pressure drop) for bundles with spiral and circumferential fins at Re from 10^4 to 10^6.

In addition to the comparison with data, we also compared our correlations with the equation by Yudin, et al. [60] for Re $> 1.8 \times 10^5$, which has the form:

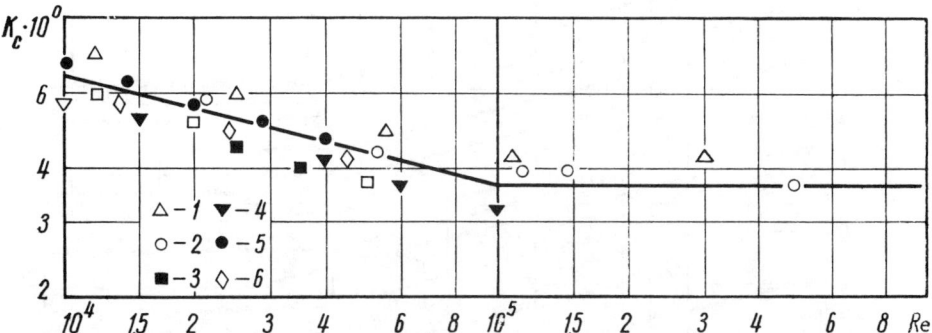

Figure 10.3 Comparison of data of other investigators with our correlations for the Euler number for staggered finned bundles. The solid curve represents our equations (5.4) and (5.5). Experimental points 1 and 2 are plotted from data of Yudin and Tokhtarova [60], points 3 and 4 are plotted from data of Yudin and Tokharova [86], points 5 are plotted from the data of Brauer [17] and points 6 from the data of Mirkovič [36]. 1) $d \times s \times h \times \delta = 23 \times 5 \times 5 \times 1$ mm, $a \times b = 3 \times 1.2$; 2) $23 \times 2.5 \times 5 \times 0.6$ mm, 2.85×1.5; 3) $32 \times 6 \times 9 \times 1.3$ mm, 2×1.2; 4) $32 \times 6 \times 9 \times 1.3$ mm, 3×1.5; 5) $28 \times 5.5 \times 5.5 \times 1.5$ mm, 1.8×1.8; 6) $25.4 \times 4.2 \times 9.5 \times 1.27$ mm, 4×2.3.

$$\xi_0 = 0.26 \left(\frac{l}{d_{eq}}\right)^{0.3} \tag{10.5}$$

We compared the four bundles used previously for thermal performance comparison. The predictions from equation 10.5 were expressed in terms of Eu as defined here and Figure 10.4 shows the results in the form Eu = f(Re) compared with the line calculated from equation (5.4) and (5.5). It is seen that the degree of agreement between the present equation and that of Yudin varies with bundle geometry. However, given the different techniques for data reduction, and also the difference in experimental conditions under which they were determined, and also the fact that Yudin's equations are suitable primarily for circumferential fins, the agreement between the data should be regarded as satisfactory.

10.3. DESIGN RECOMMENDATIONS

Recommendations on Calculating the Heat Transfer from Finned-Bundles

In determining the heat transfer coefficient for a given bundle of finned tubes, one must know the range of Re over which the exchanger is expected to perform, since the design equations for different ranges of Re are different.

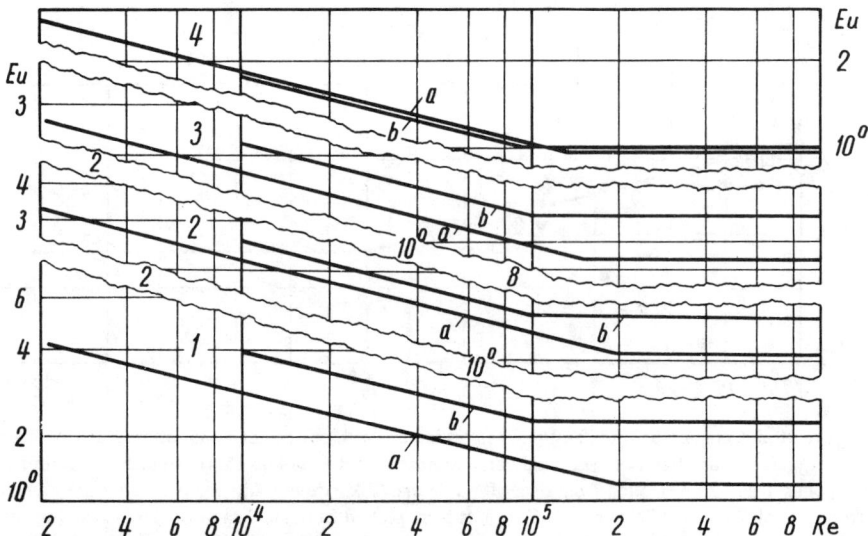

Figure 10.4 Comparison of our correlations of the Euler number for staggered bundles with equations obtained by other investigators. *a)* Yudin [60], Eqs. (10.7) and (10.8) *b)* our equations (5.4) and (5.5). For a description of bundles 1 through 4 being compared see Fig. 10.2

COMPARISON OF DATA AND PRACTICAL RECOMMENDATIONS 177

As follows from Table 8, the mixed flow zone lies in the Re range from 1×10^3 to 2×10^4; in this zone the exponent on Re for practical calculation of staggered bundles can be taken as equal to 0.65. At Re from 2×10^4 to 2×10^5, when turbulent flow starts predominating in the mixed boundary layer flow, the exponent increases to 0.8.

The critical value of Re at which transition occurs from mixed to developed turbulent boundary layer flow is close to 2×10^5. On the average, for heat transfer from bundles at $Re \geq 2 \cdot 10^5$, the exponent on Re can be taken as 0.95. Correlations for calculating the heat transfer from in-line bundles are available only for $Re \leq 4 \cdot 10^4$; these are Equations (10.2) and (10.4).

It should be noted that comparison of heat transfer from staggered and in-line finned bundles, performed on the basis of experimental data obtained by various investigators, points to the significant advantage of the staggered arrangement.

To simplify and speed up calculations on the basis of our correlations we compiled the monogram shown in Figure 10.5 which is suitable for determining the heat transfer from an interior row of a staggered bundle in the Re range

Table 8 Design recommendations on heat transfer

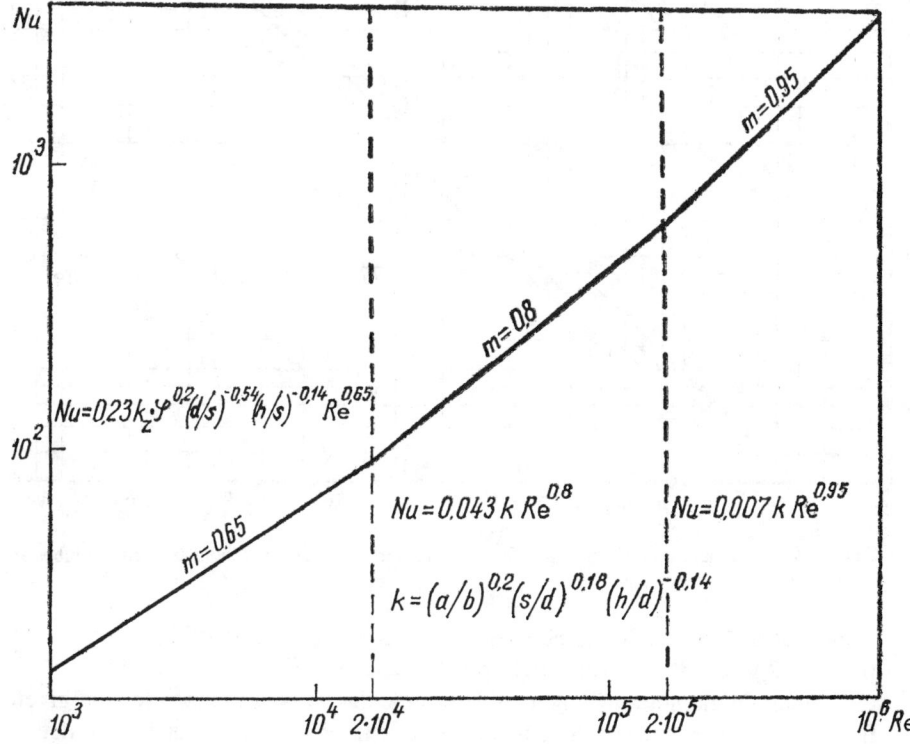

178 HEAT TRANSFER OF FINNED TUBE BUNDLES IN CROSSFLOW

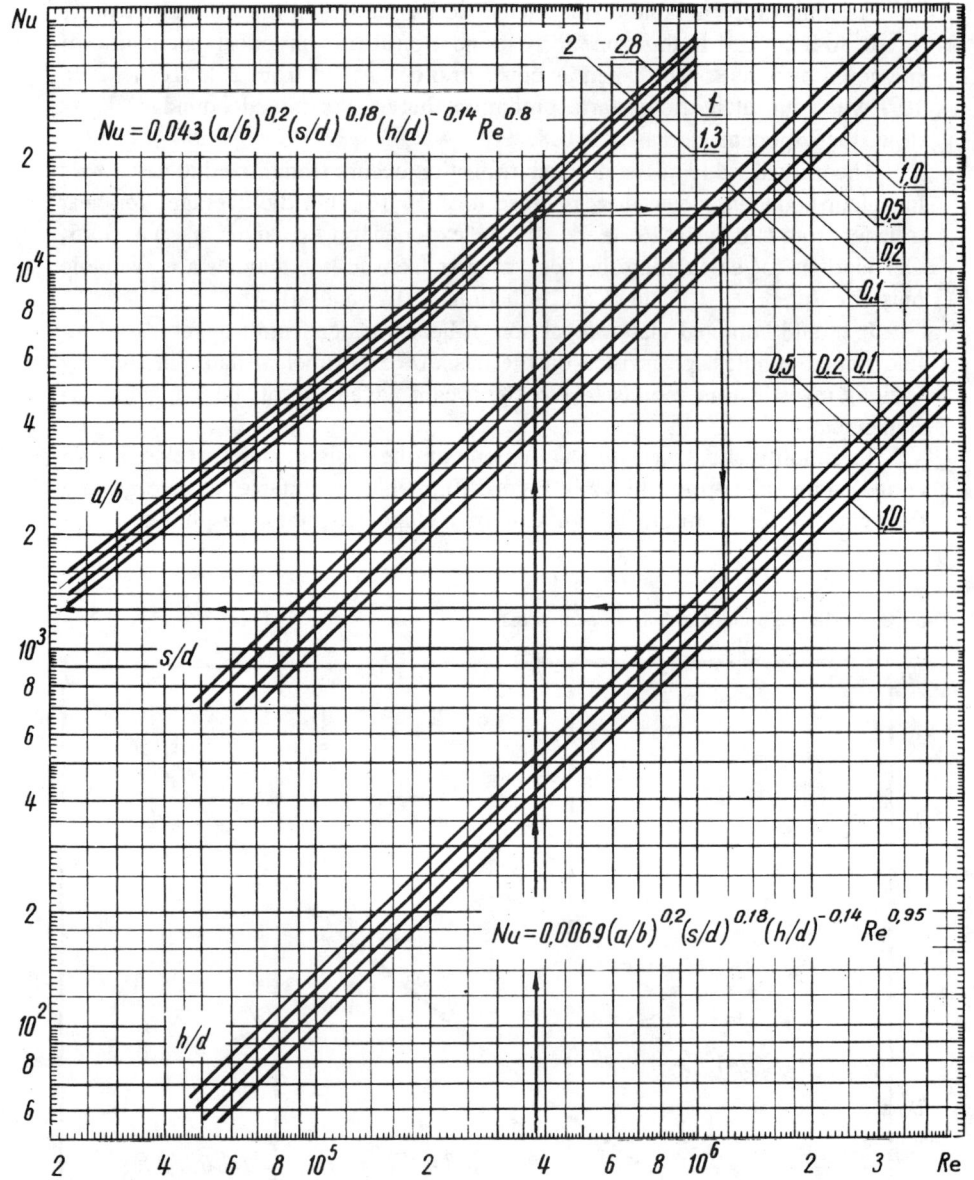

Figure 10.5 Nomogram for determining the heat transfer from staggered bundles of finned tubes at Re from $2 \cdot 10^4$ to 10^6.

from 2×10^4 to 10^6. It can be used for the following range of parameters: s/d from 0.12 to 0.28, h/d from 0.12 to 0.59, and a/b from 103 to 2.85.

Using the nomogram, one can determine the Nusselt number for the given fin and bundle geometry; the convective heat transfer coefficient is then determined from:

$$\alpha = \frac{\text{Nu} \cdot \lambda_f}{d} \tag{10.9}$$

The thermal conductivity λ_f of air should be taken at the mean coolant temperature. The reduced heat transfer coefficient can be obtained from Eq. (6.11):

$$\alpha_{\text{red}} = \left(\frac{F_{\text{fin}}}{F} E \xi \psi + \frac{F_{t\,\prime}}{F} \right) \alpha$$

The fin effectiveness E is obtained from Fig. 2.7 as a function of

$$\beta h' = \sqrt{\frac{2\alpha}{\lambda_w \delta}}$$

and d/D; here h' is the effective fin height, taking account of the effect on E of heat transfer from the fin tip. h' is determined by conditionally increasing the fin height by one half of its thickness. Thus, for a trapezoidal cross section fin

$$h' = h + \frac{\delta_1 + \delta_2}{2} \tag{10.10}$$

The coefficient ξ, correcting for the trapezoidal shape of the fin, is obtained from Fig. 3.3 for given

$$\beta h = \sqrt{\frac{2\alpha}{\lambda_w \delta}}$$

and

$$\sqrt{\frac{\delta_2}{\delta_1}}.$$

The factor ψ, accounting (within the range of βh from 0.3 to 3) for the nonuniformity of heat transfer distribution over the fin is determined from Fig. 6.11, or from expression (6.12):

$$\psi = 0.97 - 0.058\,\beta h$$

Once the reduced heat transfer coefficient is known, the value of k can be determined from Eqs. (3.18) or (3.19).

The above equations relate to an *internal* tube within the bundles. The heat transfer coefficient for the first rows of a bundle can be estimated from experimental data presented in Sec. 7.1 (Figs. 7.1 through 7.11) or from Tables 15 through 19 given in the Appendix. The nomogram shown in Figure 10.5 was compiled from formulae relating to heat transfer from a tube of an interior row in a multirow staggered bundle. When the number of rows in the flow direction is small, the average heat transfer coefficient from the bundle will be lower, this can be taken into account by using factor k_z, which may be determined from Fig. 3.12

The above equations apply to air only. For other fluids, Eqs. (7.4) and (7.5) should be multiplied by the factor $Pr^{0.4}$. Assuming that $Pr = 0.701$, for air, we can write the correlations for Re from 2×10^4 to 2×10^5 in the form:

$$Nu = 0.05 \, (a/b)^{0.2} \, (s/d)^{0.18} \, (h/d)^{-0.14} \, Re^{0.8} \, Pr^{0.4} \qquad (10.11)$$

and for $Re > 2 \times 10^5$ in the form:

$$Nu = 0.008 \, (a/b)^{0.2} \, (s/d)^{0.18} \, (h/d)^{-0.14} \, Re^{0.95} \, Pr^{0.4} \qquad (10.12)$$

2. Recommendations on Calculation of Pressure Drop for Finned Bundles

It has been seen from the above discussion of correlation that particular attention in calculations should be paid to the range of Re over which the heat exchanger will operate, since Re characterizes the pattern of flow over the bundle. The Euler number of a bundle is described by different relationships, depending on the flow mode over it.

At critical Re the initial mixed flow becomes turbulent. On the average, for the bundles under study, it was found that $Re_{cr} \simeq 10^5$. For $Re < 10^5$ the drag coefficient per transverse row ξ_0 for staggered bundles with spiral fins, can be obtained from the expression

$$\xi_0 = 13.1 \, (1 - s/d)^{1.8} \, (1 - h/d)^{-1.4} \, a^{-0.55} \, b^{-0.5} \, Re^{-0.25} \qquad (10.13)$$

In developed turbulent flow at $Re > 10^5$, the drag coefficient for staggered bundles can be calculated from the expression:

$$\xi_0 = 0.74 \, (1 - s/d)^{1.8} \, (1 - h/d)^{-1.4} \, a^{-0.55} \, b^{-0.5} \qquad (10.14)$$

The total pressure drop for a staggered multirow bundle with spiral fins can be calculated (for Re from 10^4 to 10^6 from the equation:

$$\Delta p = \xi_0 \, \frac{\rho w^2}{2} \cdot z \qquad (10.15)$$

The velocity w used in these calculations is the maximum velocity in the narrowest bundle cross section over the transverse or diagonal pitch, as appropriate. The value of Re is also calculated from the maximum velocity and external diameter of the tube carrying the fins (i.e., the fin root diameter). The coolant density ρ is determined from the mean flow temperature in the bundle. For

COMPARISON OF DATA AND PRACTICAL RECOMMENDATIONS 181

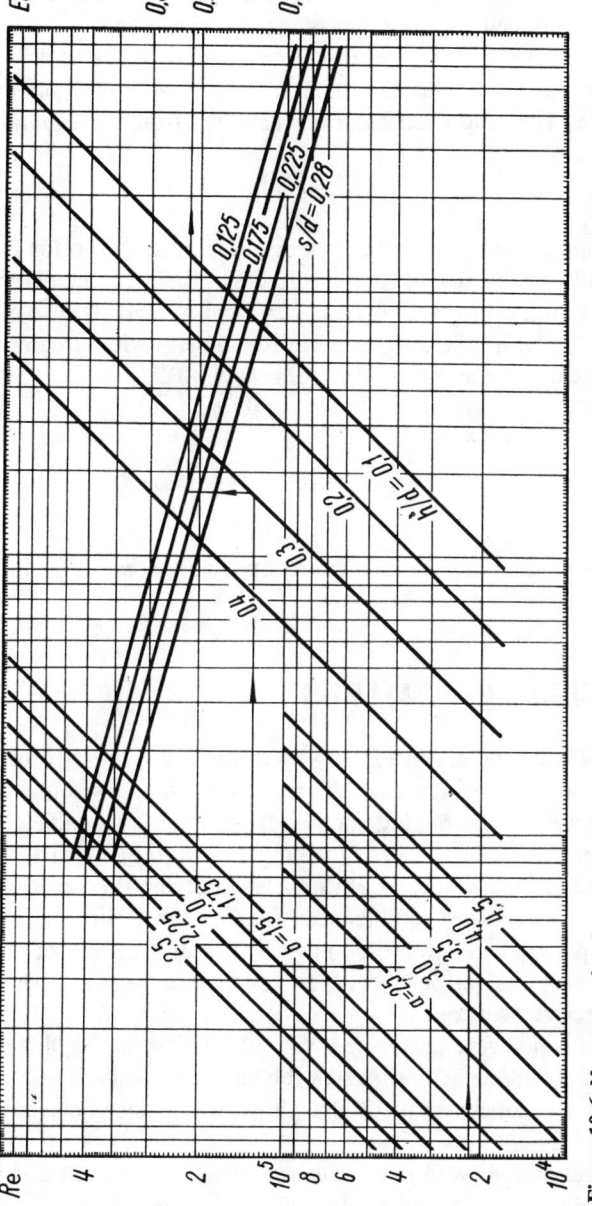

Figure 10.6 Nomogram for determining the pressure drop for flow over staggered finned tube bundles at Re from 2×10^4 to 10^6

convenience in calculating hydraulic drag with Eqs. (5.4) and (5.5) at Re from 10^4 to 10^6, we compiled a nomogram, given by Fig. 10.6.

Using the nomogram, we determine, on the basis of geometric parameters of a given bundle, the values of Eu, corresponding to one longitudinal row within a bundle. The drag coefficient is calculated from the equation

$$\xi_0 = 2\mathrm{Eu} \qquad (10.16)$$

This nomogram can be used for calculating the pressure drop for flow through staggered bundles with spiral and circumferential fins with the parameter range s/d from 0.12 to 0.28, h/d from 0.12 to 0.59 and a from 2.17 to 4.13, b from 1.27 to 2.14. The hydraulic drag of an in-line bundle with circumferential fins can be calculated for the range Re from 4×10^3 to 1.5×10^5 from the expression:

$$\xi_0 = 0.52 \left(\frac{l}{d_{eq}}\right)^{0.3} \psi^{0.68} \mathrm{Re}^{-0.08}$$

No suitable correlations for in-line finned bundles are available for Re > 1.5×10^5.

10.4. CONCLUDING REMARKS

The present study has been concerned with the analysis of experimental data on heat transfer and pressure drop for air flow over staggered bundles of finned tubes, as a function of the finning and bundle geometries, bundle configuration and flow mode. The studies covered the range of Re from 10^4 to 1.3×10^6, and were performed at the Institute of Physical Engineering Problems of Energetics of the Lithuanian Academy of Sciences. At the same time, this monograph considers the principal relationships governing thermal conduction within the fins, and also discusses various facts of the interaction between thermal conduction and convective heat transfer.

The experiments were performed with 12 bundles having different finning geometries and 9 bundles with different tube arrangements. These experiments made possible a detailed study of the physical and technical aspects of the processes.

Our experiments showed that the heat transfer coefficient and Euler number for a bundle of finned tubes depends to a large degree on the finning geometry—i.e., the pitch and height of the fin—as well as the transverse and longitudinal bundle pitches. This is expressed by different values of constants and exponents in expressions for $\mathrm{Nu}_{red} = f(\mathrm{Re})$ and $\mathrm{Eu} = f(\mathrm{Re})$ (see Tables 3 through 5).

The relationship between heat transfer coefficient and flow velocity changes significantly, depending on the mode of flow over the tube (mixed or turbulent). The flow mode is reflected in the flow pattern of the boundary layer in the leading part of the tube up to the point of flow separation, and on the nature of vortex generation in the trailing part. The change in flow mode causes changes in the exponent m on Re in relationship for the the convective heat transfer coefficient.

According to Yudin, et. al. [38], for mixed flow over staggered bundles at Re from 10^3 to 2×10^4, the value of m can be taken to be 0.65 with accuracy sufficient for practical purposes. As the value of Re is increased further, with an attendant rise in turbulence, the value of m increases. Thus, for the range Re from 10^4 to 2×10^4 it rises from 0.65 to 0.8, at Re $\simeq 2 \cdot 10^5$, developed flow turbulence sets in and m becomes equal to 0.95.

Correlations for mean heat transfer from a finned tube in an in-line bundle are available only for mixed flow over the bundle. According to Yudin, et. al. [38], $m = 0.72$ for Re from 10^3 to $2.5 \cdot 10^4$.

Heat transfer from finned tubes in an interior row is higher than that from the first rows. This increase results primarily from the turbulizing effect of the upstream tubes. This effect is different in different bundles; however, within the range of parameters in our experiments it increases with increasing fin pitch and with decreasing fin height, and decreases with an increase in the longitudinal pitch of the bundle.

A great deal of attention has been given in the present study to the development of a methodology for correlating experimental data, based on the convective heat transfer coefficient for the finned tube. It was established by special experiments that the coefficient of nonuniformity of heat transfer ψ, correcting for peculiarities in distribution of the coefficient of convective heat transfer over the surface of a finned tube, is a quantity which depends on the nondimensional fin height βh. As a result of these studies we were able to refine the expression for $\alpha_{red} = f(\alpha)$ (Eq. (6.11)). The use of this expression made it possible to correlate all the data on the mean heat transfer coefficient from staggered finned bundles by the relatively simple expressions (7.4) and (7.5).

It was found in investigating the pressure drop for flow through staggered bundles that, at Re from 7×10^4 to 2×10^5, depending on the finning parameters and bundle configuration, Eu became independent of Re. When the data for Euler numbers were approximated by a single curve, it was found that the critical value of Re for the bundles under study could, on the average, be assumed to be $\sim 10^5$, with an accuracy sufficient for practical calculations.

In correlating data on Eu for staggered for subcritical conditions (Re $<$ 10^5) bundles, the exponent r on Re can on the average be taken to be -0.25. For an in-line bundle at the same Re the value of r, according to Yudin and Tokhtarova [60] should on the average be -0.08.

In the study, proper attention was paid to questions of local heat transfer. To more correctly reflect the effect of the flow pattern on the local heat transfer from a finned tube, we used the point method of model heating. The resulting graphs (Figs. 8.5 through 8.10) illustrate rather well the distribution of local heat transfer coefficients over the circumference and height of the fins.

At the same time, we compared bundles on the basis of different characteristics. We compared the energy parameters of bundles as a function of the surface extension factor, the thermal effectiveness of bundles, and also their pumping power consumption, overall dimensions and weight (Figs. 9.3 through 9.7). Comparison shows that the best performance is obtained with densely packed bundles of tubes with low fins. Comparison of data on the basis of energy parameters shows that the thermal effectiveness of a bundle is a direct function of the coolant pressure. It was established (Eq. (9.7)) that the thermal effectiveness of a bundle increases with the coolant pressure (expressed in bars) to a power of 0.8).

In spite of the relatively wide range of aspects investigated by us, many other aspects of the problem have not yet been sufficiently explored particularly for developed turbulent flow conditions. These include questions of hydrodynamics and local heat transfer, structure of boundary layers of a finned tube, etc., which await a more detailed study. Note that, up to now, no correlations have been available on heat transfer from in-line bundles at $Re > 2.5 \times 10^4$. Additionally, the available correlations for in-line bundles, suitable for $Re < 2.5 \times 10^4$, require refinement.

Finally, we may express the hope that Eqs. (5.4), (5.5), (7.4) and (7.5), obtained by us, and allowing one to calculate the pressure drop of and the heat transfer coefficient for staggered finned bundles at $Re > 2 \times 10^4$, will make a contribution to the solution of practical problems of heat transfer from finned tubes.

APPENDICES

Appendix 1 Physical properties of air
Table 1

P, bar	1.0	2.0	3.0	5.0	8.0	12.0	17.0	25.0
					\multicolumn{4}{l}{$t=20°C$; $\mu=18.1 \cdot 10^{-6}$ N·s/m²}			
ρ, kg/m³	1.19	2.38	3.57	5.95	9.52	14.28	20.20	29.70
$\nu \cdot 10^6$, m²/sec	15.06	7.60	5.07	3.04	1.90	1.26	0.89	0.61
					\multicolumn{4}{l}{$t=30°C$; $\mu=18.6 \cdot 10^{-6}$ N·s/m²}			
ρ, kg/m³	1.15	2.30	3.45	5.75	9.20	13.80	19.56	28.75
$\nu \cdot 10^6$, m²/sec	16.17	8.08	5.39	3.23	2.02	1.34	0.95	0.65
					\multicolumn{4}{l}{$t=40°C$; $\mu=19.1 \cdot 10^{-6}$ N·s/m²}			
ρ, kg/m³	1.11	2.23	3.34	5.57	8.91	13.36	18.94	27.83
$\nu \cdot 10^6$, m²/sec	17.15	8.56	5.72	3.43	2.14	1.43	1.01	0.68
					\multicolumn{4}{l}{$t=50°C$; $\mu=19.6 \cdot 10^{-6}$ N·s/m²}			
ρ, kg/m³	1.08	2.16	3.24	5.39	8.63	12.95	18.34	26.97
$\nu \cdot 10^6$, m²/sec	18.15	9.07	6.05	3.64	2.27	1.51	1.07	0.73
					\multicolumn{4}{l}{$t=60°C$; $\mu=20.1 \cdot 10^{-6}$ N·s/m²}			
ρ, kg/m³	1.04	2.09	3.13	5.22	8.36	12.55	17.77	26.10
$\nu \cdot 10^6$, m²/sec	19.25	9.62	6.42	3.95	2.40	1.60	1.13	0.77
					\multicolumn{4}{l}{$t=70°C$; $\mu=20.5 \cdot 10^{-6}$ N·s/m²}			
ρ, kg/m³	1.02	2.03	3.05	5.08	8.13	12.19	17.26	25.4
$\nu \cdot 10^6$, m²/sec	20.10	10.10	6.72	4.04	2.52	1.68	1.19	0.81
					\multicolumn{4}{l}{$t=80°C$; $\mu=21.1 \cdot 10^{-6}$ N·s/m²}			
ρ, kg/m³	0.98	1.97	2.96	4.93	7.89	11.85	16.28	24.70
$\nu \cdot 10^6$, m²/sec	21.40	10.84	7.22	4.34	2.71	1.81	1.31	0.86
					\multicolumn{4}{l}{$t=100°C$; $\mu=21.9 \cdot 10^{-6}$ N·s/m²}			
ρ, kg/m³	0.93	1.86	2.80	4.67	7.47	11.2	15.88	23.35
$\nu \cdot 10^6$, m²/sec	23.50	11.73	7.82	4.70	2.93	1.95	1.38	0.93

Appendix 2 Experimental data on the hydraulic drag of staggered bundles of tubes with spiral fins in crossflow air

Table 2   **Bundle** $a \times b = 2.2 \times 1.3$

t_f, °C	w, m/sec	P, bar	$\Delta p \cdot 10^{-2}$ N/m²	$Re \cdot 10^{-4}$	Eu	K_c
						$d \times s \times h = 32 \times 4 \times 4$
32.2	33.7	1.17	26.0	7.6	1.75	0.464
31.2	33.6	1.27	24.9	8.32	1.52	0.404
33.1	34.1	1.71	34.0	11.2	1.52	0.404
33.7	34.0	1.76	35.9	11.5	1.57	0.417
36.6	35.0	2.16	42.8	14.2	1.47	0.390
35.7	35.0	2.33	44.7	15.4	1.40	0.371
43.9	35.7	2.65	50.5	17.1	1.38	0.37
39.2	34.3	3.04	57.1	19.3	1.44	0.38
39.7	34.6	3.14	58.6	20.1	1.42	0.38
44.8	34.2	3.87	71.7	23.8	1.45	0.38
42.5	34.6	4.12	73.1	26.0	1.36	0.36
48.1	34.2	5.15	94.3	31.0	1.45	0.38
46.1	34.1	6.00	105.6	36.7	1.39	0.37
48.8	34.0	6.96	123.5	41.5	1.43	0.38
49.8	34.6	7.75	138.8	47.1	1.40	0.37
55.5	33.7	9.71	163.3	55.2	1.41	0.37
54.1	33.2	12.6	200.9	71.8	1.37	0.36
62.2	33.2	14.0	217.8	76.0	1.36	0.37
66.0	34.0	21.1	324.1	114	1.3	0.34
69.9	34.0	24.0	362.6	126	1.29	0.34
						$d \times s \times h = 32 \times 4 \times 6$
33.5	33.0	1.29	29.9	8.28	1.86	0.446
32.2	23.1	1.29	30.1	8.31	1.86	0.446
34.7	33.8	1.94	42.8	12.5	1.7	0.407
36.6	34.5	1.94	42.3	12.7	1.62	0.387
37.8	34.1	2.67	58.3	17.1	1.68	0.401
39.4	34.4	2.68	56.2	17.1	1.59	0.380
41.4	34.1	3.39	69.1	21.5	1.57	0.38
40.2	34.9	3.57	75.7	23.1	1.56	0.37
41.8	34.0	4.51	90.3	78.2	1.56	0.37
44.0	33.5	6.35	127.5	33.7	1.62	0.39
47.9	34.1	5.88	116.2	35.6	1.56	0.37
42.6	33.3	7.94	152.7	48.4	1.57	0.38
49.4	33.3	9.90	186.0	58.1	1.57	0.38
49.6	33.0	11.2	204.5	66.2	1.59	0.36
54.4	32.8	15.3	264.3	85.9	1.51	0.36
54.8	34.2	15.3	288.2	89.6	1.51	0.36

Table 2 (*Continued*)

t_f, °C	m/sec	P, bar	$\Delta p \cdot 10^{-2}$, N/m²	$Re \cdot 10^{-4}$	Eu	K_c
66.2	34.3	24.03	419.8	131	1.44	0.34
55.4	33.9	23.53	413.1	134	1.44	0.34

$d \times s \times h = 32 \times 4 \times 9$

36.1	11.2	1.03	4.8	2.21	3.28	0.686
34.8	31.6	1.06	26.1	6.32	2.2	0.460
35.2	32.7	1.31	31.0	8.02	2.0	0.418
35.2	33.2	1.78	38.5	11.4	1.74	0.364
36.0	33.6	2.07	55.1	17.1	1.62	0.34
43.3	15.0	9.22	41.2	24.9	1.80	0.38
43.0	21.8	9.51	82.4	37.4	1.66	0.35
43.6	30.1	9.8	158.0	53.3	1.61	0.34
44.0	32.8	9.12	175.3	53.8	1.63	0.34
46.7	31.8	14.2	248.4	80.6	1.59	0.33
47.8	31.5	17.3	294.8	94.3	1.57	0.33
45.2	30.9	21.1	340.0	116.0	1.54	0.32

$d \times s \times h = 32 \times 6 \times 6$

33.4	11.6	1.05	3.23	2.4	1.99	
33.2	16.9	1.06	6.17	3.5	1.8	
33.2	23.9	1.06	10.5	4.9	1.53	
33.4	29.4	1.06	14.4	6.0	1.38	
34.5	29.9	1.48	19.9	8.3	1.31	
35.2	31.1	1.78	22.5	10.3	1.2	0.34
35.6	30.7	2.33	27.8	13.4	1.16	0.33
36.2	30.7	3.03	35.2	17.1	1.14	0.32
39.5	30.9	4.23	51.8	23.6	1.21	0.34
40.9	30.2	6.03	71.6	33.1	1.22	0.35
42.5	29.9	8.03	92.8	41.8	1.23	0.35
44.1	29.8	11.3	126	57.6	1.23	0.35
45.6	28.9	14.6	171	72.8	1.24	0.35
47.7	28.7	18.9	195	92.9	1.2	0.34
47.8	29.4	24.6	250	124.1	1.17	0.33

$d \times s \times h = 32 \times 6 \times 9$

31.2	9.6	1.04	3.43	2.04	2.1	0.49
33.0	10.3	1.04	3.43	2.13	2.75	0.64
32.7	11.4	1.03	3.63	2.30	2.39	0.56
34.7	16.9	1.05	7.65	3.67	2.17	0.50
30.4	17.4	1.05	7.65	3.73	2.1	0.49
32.0	25.3	1.05	12.9	5.2	1.69	0.39
34.5	24.9	1.05	14.0	5.4	1.84	0.43
32.2	27.5	1.06	15.2	5.83	1.68	0.39

Table 2 (*Continued*)

t_f, °C	w, m/sec	P, bar	$\Delta p \cdot 10^{-2}$, N/m^2	$Re \cdot 10^{-4}$	Eu	K_c
32.1	30.4	1.05	17.6	6.30	1.58	0.35
33.3	32.0	1.06	18.5	6.77	1.51	0.35
35.0	31.0	1.35	21.5	7.9	1.52	0.35
37.8	31.7	1.40	22.1	8.43	1.44	0.35
36.7	31.3	1.75	27.1	9.9	1.46	0.35
34.6	32.5	1.76	28.8	11.3	1.37	0.33
37.0	30.6	2.3	34.3	12.7	1.47	0.36
39.4	31.4	2.25	32.2	13.4	1.33	0.32
44.8	31.4	2.0	47.8	14.4	1.76	0.42
37.4	31.1	2.64	38.8	15.6	1.39	0.34
38.5	32.5	3.0	46.3	16.7	1.43	0.34
36.0	31.4	2.78	42.9	17.1	1.39	0.34
40.8	31.2	3.45	49.7	20.2	1.36	0.33
45.8	31.3	4.2	66.4	23.0	1.55	0.37
40.8	31.6	4.2	66.0	23.5	1.48	0.36
39.1	31.0	4.17	61.0	24.3	1.40	0.34
34.7	30.5	5.35	62.4	26.3	1.36	0.33
42.8	30.9	5.2	74.4	29.7	1.40	0.34
38.9	31.3	5.1	83.7	30.8	1.51	0.36
46.6	31.1	5.9	89.0	31.9	1.5	0.36
41.2	31.0	6.0	90.4	33.0	1.47	0.36
35.9	30.4	6.9	87.6	35.0	1.44	0.35
42.5	33.2	8.0	115.6	50.1	2.45	0.35
45.0	30.8	8.1	111.5	43.5	1.39	0.34
43.1	30.4	7.9	110.3	44.1	1.41	0.34
38.3	29.9	9.04	116.8	46.8	1.45	0.35
41.3	30.9	9.22	144.7	54.1	1.49	0.36
48.4	30.6	11.3	158.0	58.0	1.44	0.35
44.5	29.8	11.2	147.1	61.0	1.37	0.33
40.4	30.7	11.6	162.0	64.1	1.42	0.34
48.6	30.3	14.6	199.2	73.6	1.44	0.35
46.5	29.9	15.0	187.3	79.6	1.32	0.32
44.8	30.5	15.6	211.4	82.5	1.42	0.34
44.4	30.4	16.2	235.0	91.0	1.44	0.35
64.0	31.2	19.4	265.6	93.1	1.42	0.34
45.8	28.7	18.7	217.8	96.3	1.33	0.32
46.9	30.2	20.2	266.9	106.0	1.41	0.34
50.5	28.3	24.7	290.9	112.6	1.42	0.34
63.2	31.6	24.5	321.5	114.7	1.39	0.34
52.4	30.4	23.5	296.2	119.0	1.31	0.32
50.5	30.7	24.7	333.3	120.1	1.4	0.34
50.0	30.1	24.0	332.1	130.0	1.42	0.34
50.3	30.0	25.4	322.8	130.0	1.37	0.32

Table 3

Bundle $a \times b = 2.38 \times 1.46$

t_f, °C	w, m/sec	P, bar	$\Delta p \cdot 10^{-2}$, N/m^2	Re $\cdot 10^{-4}$	Eu	K_c
					$d \times s \times h = 32 \times 4 \times 13.5$	
33.0	10.2	1.05	5.49	1.99	4.56	0 75
33.6	25.4	1.05	11.1	4.97	3.36	0.55
30.1	28.2	1.05	31.3	5.64	3.37	0.55
34.5	36.3	1.05	20.8	7.07	3.10	0.51
33.4	29.1	1.29	38.4	7.19	3.12	0.51
36.3	30.6	1.61	46.7	9.23	2.8	0.46
35.3	35.4	1.37	24.9	9.26	3.03	0.50
38.2	31.1	1.86	53.9	10.5	2.71	0.45
36.2	37.0	1.69	30.3	11.9	2.7	0.44
39.8	31.2	2.35	64.4	14.8	2.56	0.42
42.4	31.4	3.23	87.7	18.3	2.52	0.41
37.5	37.0	2.84	48.6	19.8	2.57	0.42
44.6	31.1	4.68	124.9	26.0	2.55	0.42
49.2	31.4	6.11	156.8	33.3	2.45	0.4
41.8	37.5	5.16	81.7	35.6	2.36	0.39
51.2	31.0	7.94	195.2	42.4	2.40	0.4
54.7	30.7	11.2	260.4	55.8	2.35	0.39
46.6	36.8	8.73	130.2	57.4	2.36	0.39
59.8	30.5	15.3	337.4	76.8	2.29	0.38
49.9	35.7	1.38	185.9	86.5	2.30	0.38
67.1	31.6	22.3	466.2	110	2.06	0.34
65.4	31.6	22.0	491.0	110	2.19	0.36
					$d \times s \times h = 32 \times 6 \times 13.5$	
33.6	25.5	1.06	15.9	4.99	2.05	
36.2	31.3	1.29	27.7	7.58	1.97	0.37
37.9	32.0	1.61	33.6	9.53	1.86	0.35
37.9	32.3	1.87	39.0	11.2	1.8	0.34
39.2	32.2	2.29	47.2	13.6	1.8	0.34
42.0	32.2	4.59	90.3	26.9	1.73	0.4
46.6	32.0	6.11	116.8	34.5	1.73	0.4
47.7	31.7	7.8	145.4	44.1	1.71	0.4
55.0	31.3	10.8	189.9	57.1	1.71	0,4
57.7	30.7	14.5	239.8	74.5	1.67	0.39
64.5	31.8	21.7	367.9	110	1.64	0.38

Table 4

Bundle $a \times b = 2.17 \times 1.27$

t_f, °C	w, m/sec	P, bar	$\Delta p \cdot 10^{-2}$, N/m²	$Re \cdot 10^{-4}$	Eu	K_c
					\multicolumn{2}{c}{$d \times s \times h = 32 \times 8 \times 9$}	
28.3	10.4	1.03	2.65	2.2	2.07	0.54
28.0	16.4	1.04	5.79	3.5	1.78	0.46
28.8	23.6	1.05	10.3	5.0	1.54	0.40
29.2	29.5	1.05	15.0	6.3	1.44	0.37
30.6	29.2	1.36	17.7	7.9	1.36	0.35
31.8	30.0	1.76	23.0	10.9	1.32	0.34
33.8	29.7	2.3	28.8	13.5	1.28	0.33
37.4	30.1	3.0	38.8	16.8	1.31	0.35
39.5	30.1	4.2	59.7	23.2	1.47	0.39
41.8	30.0	6.0	78.4	32.5	1.38	0.36
44.4	29.7	8.0	102.3	42.6	1.39	0.37
46.9	29.0	11.3	136.8	57.0	1.37	0.36
48.8	28.5	14.6	168.6	71.8	1.37	0.36
51.4	28.0	18.9	207.2	90.4	1.35	0.36
56.1	27.2	24.5	240.4	112	1.31	0.35

Table 5

Bundle $a \times b = 2.67 \times 1.46$

$d \times s \times h = 23 \times 4 \times 4$

t_f, °C	w, m/sec	P, bar	$\Delta p \cdot 10^{-2}$, N/m²	$Re \cdot 10^{-4}$	Eu	K_c
37.8	5.0	1.02	0.78	1.41	2.75	0.82
37.2	16.4	1.04	6.28	3.23	1.99	0.59
37.1	19.8	1.04	8.44	3.91	1.84	0.55
37.7	21.1	1.45	11.7	5.68	1.66	0.50
38.6	21.4	2.4	17.7	9.39	1.47	0.44
39.4	21.5	4.33	29.1	16.9	1.33	0.4
40.3	19.5	5.83	40.3	23.0	1.34	0.4
41.7	21.3	8.06	55.9	30.7	1.42	0.42
43.4	20.9	10.9	71.7	40.3	1.4	0.42
48.6	21.8	14.3	100.9	53.3	1.4	0.42
50.1	21.4	18.8	130.2	68.0	1.43	0.43
55.2	21.2	23.9	156.8	83.0	1.41	0.42

$d \times s \times h = 23 \times 6 \times 6$

32.6	7.18	1.04	1.47	1.06	2.38	0.79
32.2	12.8	1.05	4.07	1.98	2.04	0.68
35.4	17.6	1.06	6.87	2.58	1.84	0.61
36.8	19.6	1.05	7.46	2.81	1.64	0.54
34.2	20.4	1.05	8.63	3.05	1.73	0.57
37.6	21.7	1.35	10.3	4.0	1.44	0.48
34.2	21.7	1.8	13.4	5.36	1.41	0.47
36.4	21.8	2.28	15.2	6.79	1.25	0.41
34.8	21.2	2.8	22.0	8.18	1.54	

Table 5 (*Continued*)

t_f, °C	w, m/sec	P, bar	$\Delta p \cdot 10^{-2}$, N/m^2	$Re \cdot 10^{-4}$	Eu	K_c
43.7	21.2	3.08	17.8	8.58	1.17	0.39
44.1	21.5	4.28	26.6	12.0	1.22	0.4
38.1	21.2	5.1	35.3	14.6	1.38	
44.8	21.4	6.0	37.2	16.7	1.23	0.41
42.2	21.2	8.0	50.5	22.1	1.28	0.42
40.4	20.9	9.2	58.8	25.6	1.32	
47.0	21.5	11.3	71.7	31.0	1.26	0.42
47.6	21.6	14.5	89.0	39.1	1.21	0.4
43.6	20.0	16.2	98.3	41.9	1.39	
53.2	21.1	19.2	115.8	49.9	1.27	0.42
53.7	20.8	24.0	143.6	61.2	1.29	0.43
48.8	20.5	24.0	153.6	62.8	1.38	

$d \times s \times h = 23 \times 6 \times 9$

t_f, °C	w, m/sec	P, bar	$\Delta p \cdot 10^{-2}$, N/m^2	$Re \cdot 10^{-4}$	Eu	K_c
31.3	8.2	1.03	2.45	1.19	3.09	0.79
31.0	11.0	1.04	4.12	1.6	2.88	0.73
31.0	15.9	1.03	7.06	2.31	2.36	0.60
32.3	21.6	1.76	15.8	5.3	1.69	0.43
33.0	22.0	2.28	19.3	6.94	1.55	0.4
34.3	22.0	3.14	26.6	9.48	1.56	0.4
36.2	21.8	4.29	35.7	12.8	1.56	0.4
38.5	21.7	6.10	50.7	17.7	1.58	0.4
41.2	21.7	7.99	66.6	22.9	1.59	0.41
43.3	21.0	11.3	86.8	32.5	1.53	0.39
44.9	21.0	14.6	106.8	39.5	1.51	0.39
46.6	20.7	19.3	136.1	50.8	1.51	0.39
51.8	21.5	24.5	180.2	65.0	1.48	0.38

$d \times s \times h = 23 \times 6.5 \times 10$

t_f, °C	w, m/sec	P, bar	$\Delta p \cdot 10^{-2}$, N/m^2	$Re \cdot 10^{-4}$	Eu	K_c
31.9	8.18	1.06	2.25	1.21	2.79	0.68
23.3	12.0	1.07	4.8	1.79	2.73	0.66
34.0	17.0	1.07	8.23	2.51	2.36	0.57
34.9	20.6	1.08	11.1	3.04	2.16	0.52
36.6	20.9	2.38	13.6	3.90	2.02	0.49
38.2	22.0	2.79	17.0	5.27	1.77	0.43
42.6	22.0	3.31	20.4	6.65	1.67	0.41
40.9	22.0	4.11	27.0	9.22	1.64	0.37
44.4	22.0	5.32	37.2	12.3	1.63	0.37
49.8	21.7	7.07	51.4	16.5	1.68	0.38
42.4	21.5	9.02	69.3	22.5	1.7	0.39
43.0	21.4	12.3	94.6	31.4	1.67	0.38
42.8	21.0	15.6	116.7	39.7	1.66	0.38
43.2	20.3	20.4	149.1	51.1	1.66	0.38

Table 6

Bundle $a \times b = 2.97 \times 1.65$

t_f, °C	w, m/sec	P, bar	$\Delta p \cdot 10^{-2}$, N/m²	$Re \cdot 10^{-2}$	Eu	K_c
					\multicolumn{2}{r}{$d \times s \times h = 23 \times 5 \times 13.5$}	
35.1	8.96	1.04	4.32	1.29	4.57	0.79
34.8	16.1	1.06	9.81	2.33	3.20	0.55
35.1	26.4	1.06	21.8	3.81	2.62	0.45
36.5	27.8	1.88	35.1	7.1	2.15	0.37
36.9	27.4	2.27	41.0	8.4	2.14	0.37
37.8	26.9	3.23	55.6	11.7	2.13	0.37
39.3	26.7	4.27	82.4	15.2	2.43	0.42
42.5	26.2	7.94	146.1	27.1	2.44	0.42
44.4	25.1	14.5	249.1	46.9	2.5	0.43
49.8	24.1	23.8	366.6	70.5	2.47	0.43

Table 7

Bundle $a \times b = 2.74 \times 1.64$

$d \times s \times h = 23 \times 6.5 \times 10$

t_f, °C	w, m/sec	P, bar	$\Delta p \cdot 10^{-2}$, N/m²	$Re \cdot 10^{-2}$	Eu	K_c
39.4	12.1	1.04	3.43	1.6	2.07	0.54
39.5	16.7	1.05	5.78	2.5	1.79	0.47
39.7	20.3	1.05	8.53	2.9	1.78	0.46
41.8	22.4	1.37	11.8	3.9	1.61	0.42
42.6	22.7	1.7	14.1	5.0	1.51	0.39
43.3	21.9	2.3	16.8	6.5	1.43	0.37
43.8	21.9	3.0	21.3	8.4	1.39	0.36
46.4	21.8	4.2	29.2	11.6	1.47	0.38
49.0	22.0	5.9	38.5	15.8	1.13	0.30
53.2	22.0	7.9	53.1	20.5	1.36	0.35
56.5	20.2	11.1	72.3	28.0	1.57	0.41
58.8	21.2	14.3	88.9	35.2	1.37	0.36
62.4	20.8	19.2	115.4	44.6	1.4	0.36
66.3	21.6	25.5	163.2	61.0	1.4	0.36

Table 8

Bundle $a \times b = 2.74 \times 2.11$

$d \times s \times h = 23 \times 6.5 \times 10$

t_f, °C	w, m/sec	P, bar	$\Delta p \cdot 10^{-2}$, N/m²	$Re \cdot 10^{-2}$	Eu	K_c
36.8	11.4	1.06	3.04	1.7	2.09	0.62
36.8	15.5	1.06	4.9	2.2	1.74	0.51
36.9	18.8	1.06	6.7	2.7	1.61	0.48
37.2	21.6	1.06	8.6	3.1	1.56	0.46
38.6	22.6	1.5	11.2	4.3	1.4	0.41
39.7	24.0	1.7	13.2	5.3	1.25	0.37
40.7	22.3	2.4	17.2	6.9	1.35	0.4
42.4	22.0	3.1	19.9	8.8	1.24	0.36
45.1	22.2	4.3	26.6	11.6	1.19	0.35
48.9	23.0	6.0	33.1	14.1	1.21	0.36

Table 8 (Continued)

t_f, °C	w, m/sec	P, bar	$\Delta p \cdot 10^{-2}$, N/m^2	$Re \cdot 10^{-4}$	Ru	K_c
51.6	22.1	7.9	47.7	20.5	1.22	0.36
54.6	21.6	11.4	63.7	28.9	1.17	0.34
57.4	21.6	14.6	81.0	36.1	1.18	0.35
59.0	20.7	19.2	101	44.9	1.22	0.36
61.4	20.4	24.4	126	55.5	1.24	0.36
61.2	21.7	24.7	145	60.0	1.24	0.36

Table 9

Bundle $a \times b = 3.13 \times 1.78$
$d \times s \times h = 23 \times 6.5 \times 10$

t_f, °C	w, m/sec	P, bar	$\Delta p \cdot 10^{-2}$, N/m^2	$Re \cdot 10^{-4}$	Ru	K_c
30.0	8.45	1.01	1.76	1.22	2.12	0.65
31.4	9.66	1.01	2.35	1.4	2.14	0.66
30.9	12.2	1.01	3.33	1.76	1.91	0.58
38.9	14.1	1.02	4.41	1.94	1.93	0.59
31.4	17.3	1.02	5.88	2.49	1.67	0.51
39.0	20.3	1.03	7.84	2.8	1.66	0.51
31.8	19.9	1.02	7.42	2.9	1.65	0.51
31.7	22.9	1.03	9.41	3.32	1.52	0.46
39.2	24.9	1.03	11.2	3.45	1.56	0.48
32.6	24.5	1.03	11.0	3.57	1.53	0.47
33.3	24.9	1.46	13.3	4.68	1.34	0.41
33.9	25.5	1.7	16.1	5.51	1.34	0.41
40.9	25.7	1.86	17.2	5.82	1.32	0.40
35.4	25.8	2.12	19.5	6.88	1.27	0.37
41.9	25.6	2.36	21.2	7.3	1.3	0.38
37.4	25.5	2.36	19.9	7.5	1.2	0.35
37.4	25.6	2.69	23.5	8.4	1.26	0.36
45.5	25.7	3.08	29.2	9.3	1.37	0.4
39.4	25.6	3.17	26.5	10.0	1.2	0.35
44.8	25.7	4.3	38.5	13.2	1.29	0.37
41.8	25.8	4.3	35.8	13.5	1.18	0.34
46.7	25.6	6.0	54.4	18.0	1.32	0.38
43.0	25.9	6.0	51.7	19.2	1.22	0.35
48.3	25.5	8.0	70.3	23.7	1.3	0.38
44.8	25.1	8.15	69.0	24.3	1.28	0.37
50.8	25.0	11.6	69.9	33.3	1.29	0.37
47.7	25.1	11.6	99.5	33.9	1.31	0.38
52.1	24.5	14.7	119.4	50.8	1.32	0.38

Table 10

Bundle $a \times b = 3.74 \times 2.01$

t_f, °C	w, m/sec	P, bar	$\Delta p \cdot 10^{-2}$, N/m²	$Re \cdot 10^{-4}$	Eu	K_c
					\multicolumn{2}{l}{$d \times s \times h = 23 \times 6.5 \times 10$}	
35.6	10.6	1.03	2.25	1.5	1.71	0.59
35.8	15.0	1.04	4.02	2.1	1.53	0.53
36.2	18.5	1.04	5.78	2.6	1.45	0.50
37.1	20.4	1.36	8.33	3.6	1.35	0.46
38.1	20.6	1.76	10.3	4.7	1.28	0.44
39.7	20.4	2.3	12.0	6.0	1.18	0.41
41.4	20.6	3.0	15.3	7.8	1.14	0.39
43.4	20.3	4.2	21.3	10.8	1.16	0.4
45.7	19.4	6.45	30.5	15.4	1.21	0.42
48.8	19.9	8.2	39.1	19.5	1.17	0.4
54.0	19.9	11.0	51.7	26.0	1.16	0.4
56.6	19.4	14.1	62.3	31.4	1.16	0.4
59.3	18.9	19.0	76.9	41.2	1.13	0.39
68.6	18.6	24.5	92.2	49.0	1.12	0.38
69.4	19.8	24.9	101.5	53.4	1.07	0.37

Table 11

Bundle $a \times b = 4.04 \times 1.73$

$d \times s \times h = 23 \times 6.5 \times 10$

t_f, °C	w, m/sec	P, bar	$\Delta p \cdot 10^{-2}$, N/m²	$Re \cdot 10^{-4}$	Eu	K_c
37.5	9.56	1.03	1.67	1.34	1.58	0.53
37.4	13.4	1.03	3.14	1.87	1.51	0.51
37.7	17.4	1.03	5.29	2.42	1.52	0.51
39.3	18.1	1.46	7.3	3.24	1.43	0.48
40.0	18.2	1.86	8.67	4.13	1.31	0.44
40.8	18.2	2.36	10.6	5.23	1.27	0.43
43.2	18.3	3.08	13.3	6.79	1.22	0.37
45.5	18.3	4.3	17.2	9.3	1.15	0.34
47.0	18.3	5.93	23.9	12.7	1.16	0.35
48.4	17.9	8.0	30.5	16.6	1.15	0.34
52.1	17.3	14.6	49.1	28.6	1.1	0.33
52.9	16.8	19.5	65.0	37.1	1.15	0.34
65.0	18.1	24.0	86.2	45.9	1.12	0.34

Table 12

Bundle $a \times b = 4.04 \times 1.98$
$d \times s \times h = 23 \times 6.5 \times 10$

t_f, °C	w, m/sec	P, bar	$\Delta p \cdot 10^{-2}$, N/m²	$Re \cdot 10^{-4}$	Eu	K_c
40.2	2.04	1.03	0.59	1.0	1.42	0.50
40.2	3.0	1.04	1.37	1.3	1.3	0.46
40.4	13.4	1.04	2.65	2.0	1.28	0.45
41.4	17.9	1.04	4.7	2.5	1.28	0.45
43.1	18.6	1.37	5.98	3.2	1.2	0.43
44.2	19.0	1.77	7.3	4.1	1.12	0.40
44.7	18.4	2.3	9.31	5.3	1.12	0.40
46.8	18.5	3.0	11.3	7.0	1.04	0.37
50.8	18.2	4.2	14.6	9.0	1.0	0.36
54.7	20.1	6.0	21.3	12.1	0.98	0.35
57.8	18.4	8.1	27.8	18.0	0.99	0.35
66.4	18.4	14.6	49.0	28.6	1.08	0.38
68.0	18.3	19.2	63.7	39.2	1.01	9.36
68.2	18.1	24.8	79.6	47.6	0.99	0.35

Table 13

Bundle $a \times b = 4.11 \times 2.1\bar{7}$
$d \times s \times h = 4.11 \times 2.14$

36.4	9.0	1.04	1.47	1.3	1.56	0.58
36.3	15.6	1.04	3.82	2.2	1.35	0.50
37.4	17.8	1.36	5.88	3.2	1.29	0.48
38.9	17.8	1.76	7.15	4.2	1.22	0.45
40.5	17.5	2.3	8.53	5.3	1.13	0.42
42.2	17.9	3.0	10.4	7.0	1.02	0.38
44.5	18.1	4.2	13.8	9.2	0.96	0.36
46.9	17.5	8.1	25.3	17.2	0.97	0.36
49.1	16.2	11.1	34.1	24.4	1.0	0.37
56.6	16.5	19.2	55.6	37.3	1.05	0.39
62.3	16.8	24.3	72.2	47.6	1.05	0.39

Table 14

Bundle $a \times b = 4.13 \times 1.46$
$d \times s \times h = 23 \times 6.5 \times 10$

37.7	10.01	1.03	1.86	1.44	1.6	0.49
37.4	13.6	1.04	3.14	1.94	1.46	0.45
37.1	17.4	1.04	5.1	2.52	1.44	0.44
36.9	17.6	1.37	6.08	3.1	1.37	0.42
36.7	16.9	1.77	7.55	4.03	1.34	0.41
37.2	17.5	2.9	11.8	6.76	1.22	0.38
52.1	17.7	4.2	15.7	9.37	1.16	0.36
59.6	17.5	5.9	20.5	11.8	1.13	0.35
67.0	17.6	7.9	26.8	15.5	1.13	0.35
69.6	17.2	11.3	37.8	20.0	1.17	0.36
72.0	16.9	14.2	46.1	24.6	1.17	0.36
74.0	16.1	18.7	55.1	30.9	1.18	0.36
80.6	17.3	25.0	82.7	43.3	1.17	0.36

Appendix 3 Experimental data on the mean heat transfer from staggered bundles of tubes with spiral fins in crossflow air

Table 15

Bundle $a \times b = 2.2 \times 1.3$

t_f, °C	t_w, °C	w, m/sec	P, bar	α_{red}, W/m² · K	Nu_{red}	$Re \cdot 10^{-4}$	Nu
							$d \times s \times h = 32 \times 4 \times 4$
First row							
32.1	135.4	33.7	1.17	160	202	7.6	
33.7	111.3	34.0	1.76	208	260	11.5	
36.6	137.7	35.0	2.16	244	286	14.2	
43.9	113.5	35.7	2.65	275	315	17.1	
39.7	125.3	34.6	3.14	321	371	20.1	
42.4	133.7	34.6	4.12	403	459	26.0	
46.1	133.7	34.1	6.0	530	606	36.7	
49.8	119.6	34.6	7.75	656	743	47.1	
54.0	131.0	33.2	12.5	962	1082	71.8	
66.0	116.4	34.0	21.1	1495	1635	114.0	
Fifth row							
31.2	118.0	33.6	1.27	284	349	8.32	368
33.1	145.7	34.1	1.71	367	432	11.2	474
35.7	126.9	35.0	2.33	453	531	15.4	588
39.2	129.7	34.3	3.04	546	632	19.3	710
44.7	127.7	34.2	3.87	635	726	23.8	831
48.1	122.1	34.2	5.15	786	893	31.0	1041
48.7	118.0	34.0	6.96	960	1089	41.5	1306
55.5	120.5	33.7	9.71	1183	1314	55.2	1648
62.2	111.9	33.2	14.0	1542	1685	76.0	2253
69.9	102.1	34.0	24.0	2300	2463	126.0	3742
							$d \times s \times h = 32 \times 4 \times 6$
First row							
32.2	112.3	23.1	1.29	149	177	8.31	
34.8	112.2	33.8	1.94	198	236	12.5	
37.8	126.6	34.1	2.67	254	296	17.1	
40.2	123.4	34.9	3.57	325	377	23.1	
44.0	109.6	33.5	6.35	496	570	33.7	
49.4	118.7	33.3	9.9	694	784	58.1	
54.4	110.5	32.8	15.3	930	1040	85.9	
55.4	115.4	33.9	23.5	1269	1408	134.0	

Table 15 (*Continued*)

$t_f,$ °C	$t_w,$ °C	$w,$ m/sec	$P,$ bar	α_red W/m$^2 \cdot$ K	Nu$_\text{red}$	Re $\cdot 10^{-4}$	Nu
Fifth row							
33.6	109.7	33.0	1.29	256	300	8.28	355
36.6	117.3	34.5	1.94	367	428	12.7	522
39.4	126.4	34.4	2.08	421	488	17.1	605
41.4	127.9	34.1	3.39	484	574	21.5	711
41.8	112.8	34.0	4.51	595	703	28.2	907
47.9	117.5	34.1	5.88	733	831	35.6	1166
42.6	98.9	33.3	7.94	904	1042	48.4	1569
46.6	92.7	33.1	11.2	1114	1270	66.2	2147
54.8	93.2	34.2	15.3	1393	1554	89.6	2776
66.2	95.0	34.3	24.0	1850	1980	131.0	4150

$d \times s \times h = 32 \times 4 \times 9$

$t_f,$ °C	$t_w,$ °C	$w,$ m/sec	$P,$ bar	α_red W/m$^2 \cdot$ K	Nu$_\text{red}$	Re $\cdot 10^{-4}$	Nu
First row							
35.1	111.4	31.5	1.04	95	112	6.31	
35.4	112.1	33.0	1.04	117	136	6.62	
36.9	115.3	38.6	1.37	157	184	10.0	
37.6	110.8	41.9	1.73	186	217	13.7	
39.0	114.7	37.0	3.14	255	299	21.6	
40.7	114.0	37.3	4.75	358	406	32.9	
42.5	114.0	38.2	6.59	454	523	46.0	
44.5	97.8	35.6	11.2	603	691	72.1	
48.8	90.0	37.4	15.9	780	889	104.0	
Fifth row							
36.1	98.3	11.2	1.03	100	119	2.21	127
34.8	102.6	31.6	1.06	194	228	6.36	279
35.2	113.7	32.7	1.31	230	268	8.02	330
35.2	99.8	33.2	1.78	282	330	11.4	424
36.0	97.8	33.6	2.07	362	425	17.1	590
43.3	108.5	10.0	9.22	480	554	24.9	874
43.0	100.3	21.8	9.51	606	699	37.4	1250
43.6	90.0	30.1	9.8	740	852	53.3	1510
44.0	89.0	32.8	9.12	756	871	53.8	1725
46.7	79.4	31.8	14.2	1021	1165	80.6	2470
47.8	79.4	31.5	17.3	1082	1241	94.3	2810
45.2	73.8	30.9	21.1	1207	1383	116.0	3440

Table 15 (*Continued*)

t_f, °C	t_w, °C	w, m/sec	P, bar	α_{red} W/m² · K	Nu_{red}	$Re \cdot 10^{-4}$	Nu

$d \times s \times h = 32 \times 6 \times 6$

First row

30.9	106.7	13.8	1.02	84	90	2.7	
30.3	107.1	19.9	1.028	100	118	4.0	
30.0	100.9	23.2	1.036	107	138	4.7	
29.3	101.9	27.0	1.045	131	156	5.5	
27.2	102.2	30.5	1.053	146	172	6.4	
26.5	99.4	28.4	1.37	177	210	7.8	
28.4	114.6	28.6	2.01	225	268	11.4	
30.5	120.5	28.8	3.14	293	353	17.7	
33.2	124.7	28.4	5.2	395	469	28.5	
36.5	129.7	28.4	8.24	530	623	44.3	
41.6	122.2	27.4	14.2	790	912	71.9	

Fifth row

32.2	121.7	12.8	1.035	113	172	2.6	145
31.3	111.3	19.9	1.025	154	135	3.9	200
31.1	107.9	23.9	1.033	173	205	4.8	225
28.0	106.2	27.4	1.041	188	228	5.6	259
28.2	105.3	30.4	1.049	240	251	6.2	331
27.8	106.5	23.4	1.37	248	297	6.62	343
33.0	111.9	32.8	1.47	244	345	9.34	403
36.6	109.6	39.9	1.76	394	457	13.3	547
29.0	104.4	13.3	7.26	505	517	19.0	740
27.8	114.3	9.9	10.1	467	563	20.9	680
27.9	96.9	18.3	10.8	843	1014	39.0	1390
26.2	96.9	23.4	8.43	745	898	39.7	1192
28.7	87.9	22.9	10.9	981	1177	49.0	1693
37.9	88.5	30.2	11.1	1151	1342	62.6	2027
30.4	112.1	13.0	14.5	708	846	36.8	1099
30.6	108.0	16.8	14.7	862	1029	47.7	1397
31.7	100.0	23.4	14.8	1120	1371	67.0	1994
42.3	102.2	28.4	15.0	1260	1421	77.0	2250
44.9	93.7	30.9	15.1	1500	1719	85.3	2810

$d \times s \times h = 32 \times 6 \times 6$

First row

33.0	81.0	10.3	1.04	52	61	2.13	
32.2	76.8	20.7	1.05	74	80	5.13	
32.2	75.5	27.5	1.06	87	103	5.83	
33.3	74.2	32.0	1.06	94	111	6.77	

Table 15 (*Continued*)

t_f, °C	t_w, °C	w, m/sec	P, bar	α_{red} W/m² · K	Nu_{red}	$Re \cdot 10^{-4}$	Nu
34,6	70.9	32.5	1.76	164	193	11.3	
36,0	81.6	31.4	2.78	235	276	17.1	
38,8	85.8	31.3	5.1	362	422	30.8	
41,3	82.0	30.9	9.22	555	641	54.1	
44,4	81.0	30.4	16.2	814	931	51.0	
50,0	89.4	30.1	24.0	1070	1186	130.0	

Fifth row

t_f, °C	t_w, °C	w, m/sec	P, bar	α_{red} W/m² · K	Nu_{red}	$Re \cdot 10^{-4}$	Nu
31.2	61.0	9.65	1.04	83	99	2.04	110
32.7	103.0	11.4	1.03	105	125	2.3	141
30.4	57.5	17.4	1.05	123	147	3.73	170
32.2	98.4	18.2	1.04	141	167	3.7	196
32.0	112.4	25.3	1.05	169	208	5.2	267
32.1	115.0	30.4	1.05	188	223	6.3	273
36.9	73.4	31.3	1.06	215	238	6.44	316
35.0	104.6	31.0	1.35	221	260	7.9	330
36.7	109.5	31.3	1.75	258	302	9.9	339
37.0	108.2	30.6	2.3	311	365	12.7	504
44.8	97.8	31.4	2.0	329	377	14.4	528
38.5	110.2	32.5	3.0	379	442	16.7	641
34.3	70.0	30.5	4.5	404	477	21.0	714
45.8	98.5	31.3	4.2	446	510	23.0	795
40.8	112.4	31.6	4.2	473	548	23.5	875
34.6	69.6	30.5	5.35	478	564	26.3	908
46.6	100.0	31.1	5.90	559	638	31.9	1102
41.2	107.6	31.0	6,0	586	677	33.0	1198
35.9	68.2	30.4	6.88	579	681	35.0	1199
42.5	100.4	33.2	8.0	710	820	40.1	1580
45.0	95.2	30.8	8.1	684	783	43.5	1490
38.3	72.2	29.9	9.0	707	828	46.8	1593
48.4	96.8	30.6	11.3	846	962	58.0	2001
43.5	90.6	30.4	11.3	851	980	60.7	2048
40.4	75.5	30.7	11.6	878	1020	64.1	2157
63.6	125.1	31.3	14.6	907	994	70.6	2136
48.6	99.1	30.3	14.6	968	1100	73.6	2397
44.8	75.1	30.5	15.6	1076	1235	82.5	2891
64.0	116.0	21.2	19.4	1109	1214	93.1	2889
48.5	98.4	29.5	19.5	1133	1287	96.0	3091
46.9	72.2	30.2	20.2	1290	1588	106	3857
50.5	93.0	28.3	24.7	1314	1486	112.6	3946
63.2	108.2	31.6	24.5	1281	1405	114.1	3663
50.5	91.0	30.7	24.7	1373	1553	120.1	4251
50.3	79.8	30.0	25.4	1487	1684	130.0	4922

Table 16

Bundle $a \times h = 2.38 \times 1.46$

t_f, °C	t_w, °C	w, m/sec	P, bar	α_{red} W/m² · K	Nu_{red}	$Re \cdot 10^{-4}$	Nu
First row						$d \times s \times h = 32 \times 4 \times 13{,}5$	
33.5	102.8	25.4	1.05	82	98	4.97	
34,5	107.2	36.3	1.05	117	121	7.07	
36,2	125.2	37.0	1.69	139	162	11.90	
37,5	120.5	37.0	2.84	191	223	19.8	
41,8	121.5	37.5	5.16	269	312	35.6	
49,9	97.6	35.7	13.8	445	504	86.5	
Fifth row							
33.0	110.0	10.2	1.05	73	86	1.99	104
30.1	70.1	28.2	1.05	134	160	5.64	240
33.4	103.4	29.1	1.29	156	183	7.19	302
39.8	117.8	31.2	2.35	220	258	14.8	459
42.4	123.0	31.4	3.23	263	304	18.3	590
44.6	108.3	31.1	4.68	329	375	26.0	826
49.2	111.2	31.4	6.11	366	416	33.3	977
51.2	105.7	31.0	7.94	410	463	42.4	1220
54.7	100.0	30.7	11.2	490	549	55.8	1678
59.8	99.2	30.5	15.3	566	628	76.8	2090
67.1	98.7	31.6	22.3	702	763	110.0	2880
First row						$d \times s \times h = 32 \times 6 \times 13{,}5$	
31.8	104.5	23.9	1.04	93	110	4.73	
32.6	120.4	32.9	1.04	114	136	6.48	
35.0	107.8	35.3	1.78	169	198	13.0	
38.0	118.1	35.7	2.91	231	265	19.5	
42.7	121.8	35.6	5.65	328	378	36.8	
46.6	116.6	34.9	9.51	426	480	59.4	
48.8	104.2	35.6	15.3	566	643	100.0	
Fifth row							
33.6	114.8	25.5	1.06	138	163	4.99	226
36.2	116.5	31.3	1.29	180	211	7.58	323
39.2	120.4	32.2	2.29	253	297	13.6	512
40.8	118.3	32.2	3.11	306	348	18.4	695
46.6	116.2	32.0	6.11	450	497	34.5	1279
47.7	107.2	31.7	7.8	509	578	44.0	1537
55.0	103.8	31.3	10.8	584	665	57.1	1891
57.8	100.0	30.7	14.5	685	765	74.5	2460
64.5	97.4	31.8	21.7	845	965	110.0	3443

Table 17

Bundle $a \times b = 2.17 \times 1.27$

t_f, °C	t_w, °C	w, m/sec	P, bar	α_{red} W/m² · K	Nu_{red}	$Re \cdot 10^{-4}$	Nu
						\multicolumn{2}{l}{$d \times s \times h = 32 \times 3 \times 9$}	

Fifth row

28.3	64.8	10.4	1.03	111	133	2.2	151
28.0	67.4	16.4	1.04	145	174	3.5	203
28.8	69.2	23.6	1.05	175	210	5.0	240
29.2	70.6	29.5	1.05	198	237	6.3	288
30.6	72.2	29.2	1.36	238	285	7.9	352
31.8	71.6	30.0	1.76	280	333	10.3	428
33.8	73.5	29.7	2.31	331	391	13.5	520
37.4	75.4	30.1	3.0	405	474	16.8	679
39.5	81.3	30.1	4.2	510	586	23.2	919
41.8	83.8	30.0	6.0	646	746	32.5	1283
44.4	89.7	29.7	8.0	767	881	42.6	1596
46.9	92.0	29.0	11.3	955	1090	57.0	2168
48.8	93.2	28.5	14.6	1135	1290	71.8	2373
51.4	93.6	28.0	18.9	1361	1536	90.4	3554
56.1	97.8	27.2	24.5	1550	1729	111.8	4328
55.7	97.2	27.2	24.5	1558	1739	112.3	4354

Table 18

Bundle $a \times b = 2.67 \times 1.46$

$d \times s \times h = 23 \times 4 \times$

Fifth row

37.8	43.8	5.0	1.02	130	94	1.41	100
37.2	92.5	16.4	1.04	190	159	3.23	170
37.1	96.6	19.8	1.04	208	175	3.91	183
37.7	117.0	21.1	1.45	256	216	5.68	236
38.6	98.5	21.4	2.4	380	317	9.39	350
39.0	100.5	21.5	3.04	451	378	11.9	437
39.4	94.5	21.5	4.33	593	497	16.9	577
40.3	99.8	19.5	5.83	763	638	23.0	745
41.7	100.7	21.3	8.06	983	817	30.7	980
43.4	78.4	20.9	10.9	1282	1060	40.3	1320
48.6	106.6	21.4	11.0	1300	1065	40.3	1340
48.6	103.5	21.8	14.3	1501	1227	53.3	1600
50.1	94.4	21.4	18.8	1882	1532	68.0	2060
55.2	92.5	21.2	23.9	2278	1831	83.0	2640

Table 18 (Continued)

t_f, °C	t_w, °C	w, m/sec	P, bar	α_{red} W/m² · K	Nu_{red}	$Re \cdot 10^{-4}$	Nu

$d \times s \times h = 23 \times 6 \times 6$

First row

32.6	111.8	7.18	1.04	50	43	1.06	
32.2	92.4	12.8	1.05	80	68	1.98	
35.4	96.0	17.6	1.06	90	77	2.58	
34.2	93.3	20.4	1.05	106	90	3.05	
34.2	104.7	21.7	1.8	161	136	5.36	
34.8	102.0	21.2	2.8	224	190	8.18	
38.1	106.8	21.2	5.1	384	321	14.6	
40.4	95.6	20.9	9.2	496	497	25.6	
43.6	96.0	20.0	16.2	881	726	41.9	
48.8	97.4	20.5	24.0	1215	993	62.8	

Fifth row

36.8	85.6	7.32	1.04	78	66	1.01	67
36.8	96.2	11.0	1.05	109	92	1.58	93
36.8	96.8	15.8	1.05	144	121	2.26	126
36.8	87.6	19.6	1.05	164	142	2.81	147
37.6	89.3	21.7	1.35	220	185	4.0	202
38.0	95.7	21.8	1.76	277	233	5.24	252
36.4	87.4	21.8	2.28	323	272	6.79	303
43.7	91.5	21.2	3.08	450	341	8.58	386
44.1	88.2	21.5	4.28	526	434	12.0	512
44.8	88.6	21.4	6.00	716	585	16.7	727
42.2	94.5	21.2	8.00	904	749	22.1	969
47.0	96.2	21.5	11.3	1156	953	31.0	1319
47.6	94.7	21.6	14.5	1462	1193	39.1	1795

$d \times s \times h = 23 \times 6 \times 9$

First row

31.3	86.2	8.2	1.03	56	48	1.19	
31.0	90.0	11.0	1.04	68	58	1.6	
30.9	100.0	18.3	1.04	93	80	2.69	
31.2	99.8	21.1	1.43	129	111	4.23	
33.0	97.4	21.8	2.23	191	162	6.77	
34.6	107.2	21.8	3.88	374	244	11.6	
36.9	95.8	21.6	6.82	424	360	20.1	
40.1	97.4	21.3	10.1	557	464	28.4	
46.6	93.8	21.9	15.9	792	651	44.4	
50.3	91.0	21.4	24.3	995	810	64.6	

Table 18 (*Continued*)

t_f, °C	t_w, °C	w, m/sec	P, bar	α_{red} W/m² · K	Nu_{red}	$Re \cdot 10^{-4}$	Nu
Fifth row							
31.2	107	7.4	1.02	78	67	1.08	86
31.0	109	11.0	1.03	101	87	1.6	103
31.0	104	15.9	1.03	128	110	2.31	138
31.0	92	20.3	1.04	151	129	2.96	154
31.8	114	22.0	1.35	211	182	4.14	224
32.3	104	21.6	1.76	248	210	5.3	288
33.0	109	22.0	2.28	291	249	6.94	341
34.3	95	22.0	3.14	365	310	9.48	425
36.2	100	21.8	4.29	474	400	12.8	608

$d \times s \times h = 23 \times 6.5 \times 10$

t_f, °C	t_w, °C	w, m/sec	P, bar	α_{red} W/m² · K	Nu_{red}	$Re \cdot 10^{-4}$	Nu
Fifth row							
31.9	103.3	8.2	1.05	106	91	1.21	107
32.3	100.7	12.0	1.06	131	112	1.79	134
34.0	105.9	17.0	1.07	156	134	2.51	163
34.9	105.5	20.6	1.07	174	148	3.04	183
36.6	106.8	20.9	1.38	203	173	3.9	219
38.2	108.4	22.0	1.80	240	202	5.27	267
42.6	115.8	22.0	2.33	276	230	6.65	305
40.9	114.0	22.0	3.14	334	280	9.22	407
44.4	117.0	22.0	4.38	405	335	12.3	530
49.8	120.0	21.7	6.16	508	417	16.5	730
42.4	113.8	21.5	8.15	610	507	22.5	941
43.0	111.1	21.4	11.5	768	637	31.4	1290
42.8	113.6	21.0	14.8	875	726	39.7	1670
43.2	114.8	20.3	19.8	1000	825	51.1	2070

Table 19

Bundle $a \times b = 2.97 \times 1.65$

$d \times s \times h = 23 \times 6 \times 13.5$

t_f, °C	t_w, °C	w, m/sec	P, bar	α_{red} W/m² · K	Nu_{red}	$Re \cdot 10^{-4}$	Nu
First row							
35.1	86.7	9.0	1.04	62	52	1.29	
34.8	93.6	16.1	1.06	77	65	2.33	
35.1	100.8	26.4	1.06	100	85	3.81	
36.5	100.2	27.8	1.88	146	123	7.1	
36.9	104.0	27.4	2.27	164	139	8.4	
39.3	107.5	26.7	4.27	240	200	15.2	
42.5	103.5	26.2	7.94	342	282	27.1	
44.4	59.1	25.1	14.5	482	396	46.9	
49.8	92.2	24.1	23.8	606	495	70.5	

Table 19 (Continued)

t_f, °C	t_w, °C	w, m/sec	P, bar	α_{red} W/m² · K	Nu_{red}	$Re \cdot 10^{-4}$	Nu
Fifth row							
38.4	75.6	9.1	1.02	81	68	1.28	88
38.2	74.5	15.9	1.04	112	95	2.22	136
40.0	88.1	27.6	1.04	146	122	3.78	200
40.2	90.5	28.0	1.51	188	155	5.57	263
40.1	84.4	28.3	1.9	210	177	7.11	320
42.4	91.8	28.7	2.27	234	195	8.53	367
43.8	88.8	28.6	3.48	288	239	12.8	531
46.0	91.2	28.0	5.93	398	326	21.3	867
46.7	89.8	27.7	7.96	480	384	28.1	995
53.6	90.6	27.3	10.8	550	445	36.3	1455
57.5	92.5	27.9	14.5	645	516	49.0	1888
59.9	96.4	27.5	19.2	746	593	62.1	2514
57.0	87.6	27.2	24.0	840	671	78.2	3113

Table 20

Bundle $a \times b = 2.74 \times 1.64$
$d \times s \times h = 23 \times 6.5 \times 10$

t_f, °C	t_w, °C	w, m/sec	P, bar	α_{red} W/m² · K	Nu_{red}	$Re \cdot 10^{-4}$	Nu
Fifth row							
39.8	106.2	7.2	1.01	86	78	1.0	91
39.4	108.2	12.1	1.02	120	100	1.6	119
39.5	110.2	16.7	1.03	144	121	2.4	147
41.8	111.8	22.4	1.35	197	164	3.9	206
42.6	112.8	22.7	1.7	223	185	5.0	240
43.3	114.0	21.9	2.3	263	217	6.5	291
43.8	114.0	21.9	3.0	313	259	8.4	362
46.4	118.9	21.8	4.2	390	320	11.6	476
49.0	120.2	22.0	5.9	478	389	15.8	684
53.2	124.4	22.0	7.9	582	469	20.5	870
56.5	130.0	20.2	11.7	718	576	28.0	1200
58.8	132.6	21.2	14.3	828	661	35.2	1470
62.4	131.6	20.8	19.2	987	780	44.6	1915
66.3	126.8	21.6	25.2	1221	955	61.0	2660

Table 21

Bundle $a \times b = 2.74 \times 2.11$
$d \times s \times h = 23 \times 6.5 \times 10$

t_f, °C	t_w, °C	w, m/sec	P, bar	α_{red} W/m² · K	Nu_{red}	$Re \cdot 10^{-4}$	Nu
Fifth row							
36.8	105	11.4	1.03	120	102	1.7	122
36.8	104.6	15.5	1.03	133	112	2.2	133
37.2	107.7	21.6	1.03	164	138	3.1	168
38.6	108.8	22.6	1.45	198	166	4.3	210
39.7	113.4	24.0	1.7	226	189	5.3	245
40.7	109.2	22.3	2.4	272	226	6.9	307
42.4	110.4	22.0	3.1	312	263	8.8	395
45.1	116.3	22.2	4.3	384	316	11.6	473
48.9	117.6	23.0	6.0	475	387	14.1	660
51.6	122.8	22.1	7.9	556	450	20.5	826
54.6	126.4	21.6	11.4	700	564	28.9	1150
57.4	128.7	21.6	14.6	816	655	36.1	1460
59.0	126.4	20.7	19.2	962	765	44.9	1820
61.4	119.2	20.4	24.4	1121	887	55.5	2290
61.2	116.9	21.7	24.7	1156	915	60.0	2450

Table 22

Bundle $a \times b = 3.13 \times 1.78$
$d \times s \times h = 23 \times 6.5 \times 10$

Fifth row

t_f, °C	t_w, °C	w, m/sec	P, bar	α_{red} W/m² · K	Nu_{red}	$Re \cdot 10^{-4}$	Nu
31.4	100.5	9.7	1.01	106	80	1.4	94
31.6	95.5	13.8	1.02	129	110	2.01	132
31.8	95.2	19.9	1.02	159	136	2.9	167
33.3	105.4	24.9	1.46	215	183	4.68	235
36.1	108.2	25.3	1.86	256	216	5.92	289
37.4	110.3	25.5	2.36	291	244	7.5	339
39.4	112.5	25.6	3.17	357	298	10.0	436
41.8	108.9	25.8	4.3	441	366	13.5	612
43.0	105.0	25.9	6.0	541	448	19.2	809
44.8	103.6	25.1	8.2	641	528	24.3	1030
47.7	100.8	25.1	11.6	827	677	33.9	1520

Table 23

Fifth row

Bundle $a \times b = 3.74 \times 2.01$
$d \times s \times h = 23 \times 6.5 \times 10$

t_f, °C	t_w, °C	w, m/sec	P, bar	α_{red} W/m² · K	Nu_{red}	$Re \cdot 10^{-4}$	Nu
36.0	104.6	6.4	1.01	77	65	0.89	74
35.6	100.0	10.6	1.02	108	91	1.5	107
35.6	104.9	15.0	1.03	135	116	2.1	139
36.2	104.3	18.5	1.03	152	128	2.6	156
37.1	113.4	20.4	1.35	187	157	3.63	197
38.1	117.6	20.6	1.75	216	183	4.7	235
39.7	117.3	20.4	2.3	249	208	6.0	284
41.4	119.2	20.6	3.0	290	241	7.8	334
45.7	118.9	19.4	6.4	437	361	15.4	570
48.8	122.2	19.9	8.2	533	435	19.5	751
54.0	128.0	19.9	11.0	648	524	26.0	1060
56.6	127.0	19.4	14.1	745	596	31.4	1260
59.3	122.2	18.9	19.0	882	702	41.2	1620
68.6	127.0	18.6	24.5	1027	800	49.0	1970
69.4	125.1	19.8	24.9	1061	824	53.4	2100

Table 24

Fifth row

Bundle $a \times b = 4.04 \times 1.78$
$d \times s \times h = 23 \times 6.5 \times 10$

t_f, °C	t_w, °C	w, m/sec	P, bar	α_{red} W/m² · K	Nu_{red}	$Re \cdot 10^{-4}$	Nu
37.5	100.4	9.6	1.03	95	90	1.34	105
37.4	100.3	13.4	1.03	115	111	1.87	129
37.4	99.4	16.5	1.03	130	126	2.29	151
39.3	98.9	18.1	1.46	158	154	3.24	187
40.0	99.3	18.2	1.86	186	181	4.13	232
40.8	90.5	18.2	2.36	212	207	5.23	263
43.2	99.4	18.3	3.08	250	242	6.79	325
45.5	101.1	18.3	4.3	305	299	9.7	421
46.9	101.6	18.3	5.93	375	369	12.7	571
48.4	102.8	17.9	8.0	449	444	16.7	754
50.6	103.0	17.4	11.3	546	537	22.7	965
52.1	97.2	17.3	14.5	654	643	28.6	1260
52.9	102.4	16.8	19.5	758	746	37.1	1510
65.0	119.1	18.1	24.0	844	806	45.9	1820

Table 25

Fifth row

Bundle $a \times b = 4.04 \times 1.98$
$d \times s \times h = 23 \times 6.5 \times 10$

t_f, °C	t_w, °C	w, m/sec	P, bar	α_{red} W/m² · K	Nu_{red}	$Re \cdot 10^{-4}$	Nu
40.2	113.4	2.0	1.02	89	74	1.0	85
40.2	101.4	3.0	1.03	106	88	1.3	103
40.4	110.4	13.4	1.03	128	106	2.0	126
41.4	121.2	17.9	1.03	152	126	2.5	154
43.1	113.7	18.6	1.35	178	148	3.2	195
44.2	116.6	19.0	1.75	207	171	4.1	219
44.7	117.6	18.4	2.3	241	199	5.3	267

Table 25 (Continued)

t_f, °C	t_w, °C	w, m/sec	P, bar	α_{red} W/m² · K	Nu_{red}	$Re \cdot 10^{-4}$	Nu
46.8	121.5	18.5	3.0	235	234	7.0	331
50.8	120.8	18.2	4.25	348	278	9.0	397
54.7	132.6	20.1	6.0	430	351	12.1	588
57.8	123.5	18.4	8.1	517	426	18.0	750
61.3	130.7	18.9	11.3	636	504	23.2	967
66.4	124.4	18.4	14.6	761	595	28.6	1270
68.0	116.6	18.3	19.2	910	710	39.2	1680
68.2	110.8	18.1	24.8	1004	785	47.6	1940

Table 26

Bundle $a \times b = 4.11 \times 2.14$
$d \times s \times h = 23 \times 6.5 \times 10$

Fifth row

36.4	100.0	9.0	1.03	95	80	1.3	92
36.2	103.6	12.3	1.03	114	96	1.8	113
37.4	105.6	17.8	1.35	164	138	3.2	168
38.9	106.9	17.8	1.8	190	159	4.2	198
40.5	108.2	17.5	2.3	221	184	5.3	239
42.2	111.1	17.9	3.0	258	214	7.0	286
44.6	113.4	18.1	4.2	316	261	9.2	373
45.7	115.0	17.7	6.0	390	321	13.0	510
46.9	115.0	17.5	8.1	474	389	17.2	674
49.1	118.6	16.2	11.1	575	469	24.4	865
52.2	121.2	16.8	14.3	673	544	28.8	1080
56.6	123.8	16.5	19.2	801	641	37.3	1410
62.3	128.0	16.8	24.3	933	737	47.6	1710

Table 27

Bundle $a \times b = 4.13 \times 1.46$
$d \times s \times h = 23 \times 6.5 \times 10$

Fifth row

37.7	108.2	10.0	1.02	109	92	1.44	107
37.4	110.4	13.6	1.03	132	111	1.94	132
37.1	109.2	17.4	1.03	156	131	2.52	159
36.7	113.4	16.9	1.75	207	174	4.03	223
36.8	130.6	17.5	2.3	249	210	5.32	278
37.2	117.6	17.5	2.9	290	244	6.76	337
52.1	137.8	17.7	4.2	365	296	9.37	446
59.6	145.2	17.5	5.9	432	343	11.8	565
67.0	143.0	17.6	7.9	534	417	15.5	750
69.6	147.8	17.2	11.3	647	502	20.0	993
72.0	139.8	16.9	14.2	731	564	24.6	1180
74.0	132.0	16.1	18.7	854	655	30.9	1490
80.6	127.4	17.3	25.0	1057	797	43.3	2150

Appendix 4 Experimental data on local heat transfer from a finned tube with $d = 32$ mm, s from 4 to 7 mm, $h = 13.55$ mm in crossflow air

Table 28

$w = 23.5$ m/sec; $P = 1.01$ bar; Re $= 4.3 \cdot 10^4$

s	$t_f,°C$	$t_1,°C$	$t_2,°C$	$t_3,°C$	$t_4,°C$	$t_5,°C$	α_1	α_2	α_3	α_4	α_5	α_h
												$\varphi = 0°$
4	27.5	69.6	83.4	85.2	78.1	73.8	258	195	188	215	235	207
5	27.8	70.2	83.1	83.1	75.8	72.9	255	196	196	226	240	212
6	28.1	70.2	81.7	81.7	75.0	72.4	258	202	202	231	245	218
7	28.5	70.2	80.4	80.8	75.4	73.0	261	209	208	232	244	222
												$\varphi = 45°$
4	26.1	64.4	70.2	69.2	66.6	65.1	274	238	245	259	269	251
5	26.5	63.6	69.2	67.8	65.2	63.8	283	246	254	272	282	261
6	26.5	63.2	68.1	66.8	64.4	63.0	286	253	261	277	288	267
7	26.6	62.8	67.4	66.2	64.2	62.8	299	257	265	279	290	271
												$\varphi = 90°$
4	25.6	67.5	73.2	70.6	74.8	80.8	254	224	237	217	193	222
5	25.5	66.0	69.4	66.2	68.6	74.4	261	241	260	245	216	242
6	25.6	65.5	68.3	64.8	66.6	72.5	263	246	268	256	224	249
7	25.8	65.5	67.6	64.6	66.6	72.5	264	251	271	257	225	252
												$\varphi = 135°$
4	25.7	90.0	103.0	106.7	105.8	102.0	181	151	144	146	152	150
5	25.8	83.0	94.8	97.3	96.3	93.0	201	167	161	163	171	168
6	2 .7	98.6	85.0	87.7	87.5	84.7	152	187	179	179	188	182
7	25.7	76.4	86.2	89.6	89.1	86.7	214	180	170	171	178	178

Table 28 (*Continued*)

s	t_f,°C	t_1,°C	t_2,°C	t_3,°C	t_4,°C	t_5,°C	α_1	α_2	α_3	α_4	α_5	α_h
												$\varphi=180°$
4	24.2	84.0	97.7	102.0	102.2	98.5	190	154	146	146	153	153
5	24.5	70.8	82.0	86.0	86.4	83.5	236	190	178	177	185	187
6	24.7	67.2	76.8	79.5	80.1	78.0	259	211	201	199	206	209
7	25.0	70.4	80.6	83.5	84.2	81.2	243	198	188	186	196	196
												$\varphi=225°$
4	31.2	92.4	105.2	108.1	107.7	103.5	178	146	141	143	150	147
5	31.4	82.6	83.8	96.8	96.6	92.5	202	198	158	159	169	177
6	31.5	76.4	86.0	89.7	90.4	87.5	228	188	176	174	183	183
7	31.7	77.2	86.7	91.2	93.1	90.8	224	186	172	166	173	179
												$\varphi=270°$
4	24.4	66.2	74.7	76.2	81.0	87.2	264	220	214	195	176	208
5	24.6	64.3	69.6	68.6	69.2	74.0	276	243	249	245	221	242
6	25.0	63.0	67.6	66.0	65.7	70.2	286	255	269	267	240	257
7	25.4	63.0	66.6	66.0	65.7	69.7	287	262	266	268	243	261
												$\varphi=315°$
4	29.6	66.9	73.5	73.0	69.4	68.5	278	236	239	260	266	248
5	30.0	65.0	70.8	70.0	67.0	66.2	295	253	260	279	285	267
6	30.2	65.3	70.6	69.6	66.0	65.8	285	248	250	274	282	261
7	30.6	65.0	70.2	66.4	69.9	66.0	290	253	258	276	283	265

Table 29

$w = 23.4$ m/sec; $P = 3.7$ bar; $Re = 16 \cdot 10^4$

s	t_f, °C	t_1, °C	t_2, °C	t_3, °C	t_4, °C	t_5, °C	α_1	α_2	α_3	α_4	α_5	$\bar{\alpha}_h$
												$\varphi = 0$
4	34.3	58.8	65.0	66.4	65.3	65.0	590	471	451	468	471	472
5	34.5	59.1	65.7	67.4	66.1	66.6	596	470	446	464	457	467
6	34.6	59.1	66.0	67.0	65.8	66.4	599	467	453	471	461	469
7	34.8	59.4	66.2	66.2	65.4	66.4	584	457	457	468	454	464
												$\varphi = 45°$
4	33.8	51.8	52.6	54.6	57.4	58.4	783	750	682	598	574	679
5	32.4	49.6	50.2	51.7	54.0	55.0	819	791	730	650	623	724
6	33.1	49.8	50.0	51.3	53.6	54.5	843	830	772	686	657	762
7	33.4	50.8	51.1	52.5	54.8	55.4	803	789	731	651	636	727
												$\varphi = 90°$
4	39.2	56.5	53.8	55.8	67.2	79.5	855	888	892	529	367	735
5	38.2	54.4	51.8	53.4	61.0	70.2	911	1080	967	648	463	859
6	37.2	53.5	51.1	52.0	57.0	65.3	886	1036	976	727	514	865
7	36.0	52.4	50.5	51.1	55.8	63.0	882	993	956	797	533	858
												$\varphi = 135°$
4	40.8	84.0	100.2	109.0	109.2	107.6	359	261	227	226	232	247
5	40.7	78.2	91.0	99.5	102.5	101.6	435	325	278	265	268	299
6	40.9	72.6	81.5	89.3	94.7	93.6	501	391	328	295	301	350
7	41.0	75.2	84.0	94.2	97.6	97.0	463	369	298	280	283	327
												$\varphi = 180°$
4	35.7	88.0	105.0	116.2	118.6	115.0	331	249	215	209	216	233
5	36.5	74.0	84.0	90.5	93.1	91.6	433	343	302	288	295	320
6	37.3	75.9	86.2	92.2	93.3	92.2	422	333	297	291	297	316
7	38.3	77.1	88.4	93.1	94.2	92.9	420	325	297	291	302	314
												$\varphi = 225°$
4	35.3	92.2	110.0	116.6	115.4	111.3	304	226	212	216	228	226
5	35.2	83.8	101.6	107.4	105.3	100.2	343	250	231	238	256	250
6	36.1	78.0	91.2	98.2	97.9	95.8	390	297	264	265	275	285
7	36.2	74.8	86.4	95.6	97.9	94.7	422	325	274	264	278	299
												$\varphi = 270°$
4	35.8	48.7	48.7	51.6	61.5	72.1	1040	1040	852	523	370	787
5	35.8	48.6	48.0	50.0	55.8	64.8	1055	1100	954	726	466	880
6	35.8	48.7	48.3	49.8	55.1	63.4	1050	1090	969	701	490	883
7	35.8	48.6	48.3	49.8	54.4	61.8	1050	1070	959	730	519	882
												$\varphi = 315°$
4	35.1	54.6	54.6	52.8	55.8	58.6	713	713	782	669	590	700
5	35.2	54.0	54.4	52.8	55.6	57.7	735	724	789	678	616	714
6	35.3	54.0	54.6	53.1	54.6	54.0	739	720	779	720	739	741
7	35.5	54.0	54.6	53.4	56.4	57.0	752	727	773	671	645	717

Table 30

$w = 28.0$ m/sec; $P = 3.4$ bar; Re $= 18 \cdot 10^4$

s	t_f,°C	t_1,°C	t_2,°C	t_3,°C	t_4,°C	t_5,°C	α_1	α_2	α_3	α_4	α_5	$\bar{\alpha}_h$
												$\varphi = 0°$
4	36.7	61.4	66.9	68.3	69.9	70.8	821	671	642	611	594	648
5	36.7	60.9	66.2	67.2	67.6	68.3	832	683	659	651	638	671
6	36.7	60.8	66.0	66.9	66.9	67.4	831	683	662	662	651	676
7	36.7	60.4	63.6	66.2	66.0	66.4	843	669	678	683	672	683
												$\varphi = 45°$
4	36.8	53.8	64.8	57.4	63.2	64.7	1280	1210	1060	827	799	1042
5	36.8	54.0	54.8	57.0	62.3	63.0	1280	1220	1090	865	842	1067
6	36.7	54.0	54.6	66.8	62.2	62.6	1260	1230	1080	859	845	1067
7	36.7	54.4	54.6	56.6	61.6	61.8	1230	1220	1100	878	869	1075
												$\varphi = 90°$
4	36.9	54.0	54.8	56.8	69.4	73.0	1276	1215	1093	672	603	993
5	36.8	53.6	54.1	55.6	62.5	64.8	1294	1256	1156	844	776	1081
6	36.8	53.4	53.8	55.4	62.0	64.2	1310	1272	1169	860	794	1098
7	36.8	53.6	53.3	55.1	60.7	62.2	1294	1283	1187	907	856	1124
												$\varphi = 135°$
4	37.0	55.4	61.4	71.3	88.0	92.4	1210	909	648	436	401	693
5	37.0	54.6	57.0	64.2	72.8	83.1	1243	1092	806	511	511	826
6	36.9	54.2	56.2	62.4	78.4	81.7	1269	1133	860	527	490	861
7	36.9	54.1	55.4	60.2	75.0	78.7	1266	1182	936	572	521	910
												$\varphi = 180°$
4	37.0	69.6	82.2	82.7	99.2	99.8	709	513	420	372	369	452
5	37.0	67.4	77.5	86.2	92.6	93.3	753	566	467	413	407	498
6	37.0	66.0	74.5	82.2	88.5	89.1	791	611	508	445	440	538
7	37.0	65.3	72.0	79.2	87.0	88.2	818	661	549	458	448	571
												$\varphi = 225°$
4	36.8	62.0	72.0	80.8	90.2	92.8	923	681	529	435	415	562
5	36.9	61.0	65.2	66.8	80.4	83.6	941	804	761	522	486	700
6	37.0	60.8	63.6	64.0	77.8	81.4	955	855	841	556	512	750
7	37.0	60.4	61.7	60.4	74.4	78.6	972	919	972	607	546	821
												$\varphi = 270°$
4	34.2	57.0	56.0	53.8	71.0	76.6	992	1036	1149	614	532	909
5	34.8	57.4	56.7	53.6	64.8	69.0	1000	1029	1201	752	650	964
6	35.3	57.8	57.0	53.8	63.3	66.9	1000	1048	1227	814	722	998
7	35.8	57.6	57.4	53.6	61.1	63.7	1041	1050	1273	895	811	1041
												$\varphi = 315°$
4	30.6	52.4	52.8	49.6	56.0	58.8	1002	979	1147	871	771	969
5	31.6	52.8	53.4	50.0	54.8	57.0	1025	1000	1182	933	850	1013
6	32.2	53.8	53.8	50.5	55.1	57.0	1036	1036	1226	978	903	1053
7	32.8	53.8	54.2	51.1	55.6	57.4	1064	1047	1226	934	910	1060

Table 31

$w = 22.4$ m/sec; $P = 20.4$ bar; $Re = 76 \cdot 10^4$

s	t_f,°C	t_1,°C	t_2,°C	t_3,°C	t_4,°C	t_5,°C	α_1	α_2	α_3	α_4	α_5	$\bar{\alpha}_h$
												$\varphi = 0°$
4	54.4	91.4	100.2	100.4	100.2	107.8	1620	1310	1300	1310	1120	1283
5	54.5	89.2	95.5	99.4	97.5	110.8	1710	1450	1320	1380	1050	1334
6	54.7	88.9	95.4	99.0	97.5	112.3	1730	1490	1340	1380	1030	1351
7	54.8	88.6	93.3	98.2	97.5	113.4	1730	1520	1350	1370	999	1356
												$\varphi = 45°$
4	54.3	79.4	76.6	79.6	85.6	88.6	2420	2720	2400	1940	1770	2326
5	54.4	78.7	75.9	79.0	83.5	86.8	2500	2820	2470	2060	1870	2419
6	54.4	78.0	75.9	78.2	82.5	85.9	2490	2730	2430	2090	1870	2387
7	54.3	79.0	77.2	78.7	84.2	86.6	2500	2710	2530	2070	1920	2409
												$\varphi = 90°$
4	53.8	75.2	73.8	76.2	87.5	100.2	2910	3110	2790	1850	1340	2491
5	54.0	74.0	73.5	75.6	85.2	99.2	3130	3200	2890	2000	1380	2587
6	54.2	74.0	73.5	75.4	83.1	95.6	3130	3210	2920	2140	1490	2637
7	54.3	74.5	73.5	75.2	82.0	89.3	3090	3250	2990	2260	1780	2744
												$\varphi = 135°$
4	58.1	84.5	90.8	95.4	100.2	104.4	1340	1080	946	837	763	963
5	58.3	83.1	87.7	93.1	98.4	101.4	1410	1190	1010	875	814	1037
6	58.5	83.1	87.5	92.8	97.5	100.2	1440	1220	1030	908	849	1066
7	58.7	81.2	86.2	91.2	95.6	99.0	1500	1280	1090	950	869	1113
												$\varphi = 180°$
4	57.3	77.1	79.2	81.7	85.4	89.1	1160	1050	942	819	724	931
5	57.5	76.4	77.1	79.4	83.1	85.4	1260	1210	1080	930	852	1069
6	57.7	76.8	79.2	81.5	84.5	86.4	1240	1110	1000	888	830	1002
7	57.8	77.1	79.8	82.2	84.2	85.4	1230	1080	976	900	863	993
												$\varphi = 225°$
4	56.1	72.2	80.1	83.9	86.7	91.2	1380	923	797	724	631	781
5	56.2	70.2	77.1	82.8	86.6	89.3	1730	1540	906	791	728	867
6	56.3	67.2	70.6	77.1	82.8	87.4	2050	1560	1070	839	716	952
7	56.6	66.0	68.3	74.8	80.8	85.6	2360	1900	1220	920	764	1034
												$\varphi = 270°$
4	54.4	61.6	60.7	61.6	67.2	76.6	3160	3610	3160	1780	1020	2295
5	54.9	81.8	61.4	61.8	63.9	69.2	3270	3500	3270	2530	1600	2418
6	55.4	62.3	61.8	62.3	63.9	69.1	3260	3470	3260	2630	1640	2447
7	55.7	62.3	62.0	62.3	63.9	68.3	3370	3500	3370	2700	1760	2499
												$\varphi = 315°$
4	55.5	81.2	82.0	78.4	79.6	88.0	2240	2180	2510	2390	1780	2209
5	55.0	80.6	81.4	78.2	79.6	87.3	2300	2230	2520	2390	1820	2241
6	55.1	80.4	81.7	78.1	80.1	81.3	2300	2200	2540	2340	1810	2228
7	54.9	80.8	82.2	79.0	80.6	87.5	2270	2150	2440	2290	1800	2174

Appendix 5 Experimental data for determination of the coefficient of nonuniformity of heat transfer over a finned tube with $d = 32$ mm, $s = 6$ mm and $h = 9$ mm in a staggered bundle

Table 32

t_f, °C	t_w, °C	t_m, °C	P, bar	w, m/sec	α_{red} W/m²·K	$\alpha_{\ast}$ W/m²·K	$Re \cdot 10^{-4}$	βh	E	ψ
Spiral fin										
First row										
33.0	81.0	77.9	1.02	10.3	51	55	2.13	0.36	0.95	0.950
32.2	76.8	71.4	1.03	20.7	70	79	5.13	0.42	0.93	0.930
32.2	75.5	71.4	1.03	27.5	87	96	5.83	0.47	0.92	0.953
33.3	74.2	69.2	1.04	32.0	94	107	6.77	0.5	0.91	0.929
34.6	70.9	65.2	1.76	32.5	164	194	11.3	0.67	0.85	0.942
36.0	81.6	72.2	2.76	31.4	235	296	17.1	0.83	0.79	0.937
38.8	85.8	72.1	5.10	31.3	363	512	30.8	1.09	0.70	0.915
Fifth row										
31.2	60.1	57.8	1.02	9.6	83	93	2.04	0.469	0.91	0.942
30.4	57.5	54.2	1.03	17.4	122	140	3.73	0.565	0.89	0.940
30.8	61.2	57.1	1.03	22.6	131	152	4.82	0.591	0.885	0.935
31.0	70.1	63.8	1.04	30.5	163	195	6.53	0.670	0.856	0.930
32.1	64.4	58.3	1.38	31.3	197	244	8.72	0.754	0.82	0.931
33.1	67.0	59.2	1.96	30.8	249	323	12.1	0.863	0.78	0.912
33.8	74.0	63.2	2.45	31.2	305	417	15.3	0.984	0.73	0.913
35.9	68.1	55.4	5.88	30.4	579	1068	35.0	1.57	0.54	0.843
37.0	71.2	56.0	8.04	29.9	708	1275	46.8	1.72	0.51	0.895
50.3	79.8	61.6	24.4	30.0	1488	3882	130.0	3.0	0.3	0.858
32.7	103.0	96.2	1.01	11.4	105	121	2.3	0.529	0.90	0.926
32.2	98.4	90.3	1.02	18.2	141	168	3.7	0.624	0.87	0.922
32.0	112.4	100.4	1.03	25.3	169	207	5.2	0.693	0.84	0.915
32.1	115.0	101.7	1.04	30.4	188	233	6.3	0.735	0.81	0.908
35.0	104.6	92.2	1.32	31.0	221	279	7.9	0.806	0.80	0.915
36.7	109.5	93.2	1.72	31.3	257	346	9.9	0.895	0.77	0.896
37.0	108.2	91.7	2.26	30.6	311	428	12.7	0.990	0.73	0.896
38.5	110.1	89.8	2.94	32.5	379	549	16.7	1.13	0.68	0.891

Table 32 (*Continued*)

t_f, °C	t_w, °C	t_m, °C	P, bar	w, m/sec	α_{red} W/m²·K	α, W/m²·K	Re·10⁻⁴	βh	E	ψ
40.8	112.4	89.2	4.12	31.6	472	725	23.5	1.30	0.63	0.905
41.2	107.5	82.0	4.90	31.0	586	990	33.0	1.52	0.55	0.892
42.5	100.4	75.0	7.85	33.2	711	1315	46.6	1.75	0.51	0.859
43.5	90.6	68.1	11.1	30.4	852	1696	70.5	1.98	0.45	0.880
63.6	125.1	92.6	14.3	31.3	907	2070	82.2	2.19	0.40	0.815
64.0	116.0	96.1	19.0	31.2	1109	2720	108.5	2.51	0.35	0.825
63.2	108.2	81.2	24.0	31.6	1281	3350	133.6	2.79	0.32	0.815

Circumferential fin
Fifth row

24,2	92,9	78,4	1,02	10,8	121	140	2.3	0.57	0.89	0.932
24,5	100,6	76,8	1,03	15,5	173	214	3.3	0.69	0.84	0.902
24,8	101,5	74,5	1,03	22,4	205	264	4.8	0.77	0.81	0.892
28,4	96,3	69,4	1,99	30,8	383	565	11.7	1.13	0.68	0.880
31,5	86,1	59,4	3,33	31,5	568	975	19.5	1.48	0.56	0.875
32,6	77,5	52,8	4,4	31,4	715	1350	21.8	1.77	0.49	0.872

REFERENCES

1. Schmidt, E. Die Wärmeübertragung durch Rippen. Z. VDI, 1926, v. 70.
2. Eckert, E. R. G. and Drake, R. M. Heat and mass transfer. McGraw-Hill, 1959. Russian translation 1961. Re-issue, Hemisphere Publishing, 1987.
3. Makhin, V. A., Belik, N. P. and Kosarev, A. A. Concerning the calculation of heat transfer in straight fins of variable thickness. Gidroaeromekhanika (Hydroaerodynamics), 1965, no. 1.
4. Harper, D. R. and Brown, W. B. Mathematical equations of heat conduction in the fins of air-cooled engines. NACA, Rep., 1923, no. 158.
5. Bogaerts, C. and Meyer, P. Die Berechnung und Messung des Temperaturverlaufes in Wärmeübertragungsrippen. Forsch. Geb. Ing.-Wes., 1931. 2.
6. Il'in, L. N. and Styrikovich, M. A. Simplified calculation of heat transfer in circumferential fins of circular tubes. Sovetskoye kotloturbinostroyeniye, 1940, no. 2.
7. Gardner, K. A. Efficiency of extended surface. Trans. ASME, 1945, v. 67, no. 8.
8. Antuf'yev, V. M. and Beletskiy, G. S. Teploperedacha i aerodinamicheskiye soprotivleniya trubchatykh poverkhnostey v poperechnom potoke (Heat transfer and aerodynamic drag of tubular surfaces in crossflow). Mashgiz Press, Moscow, 1948.
9. Baklastov, A. M. Candidate thesis at the Moscow Energetics Institute, 1949.
10. Antuf'yev, V. M. Comparative studies of heat transfer and drag of finned surfaces. Energomashinostroyeniye, 1961, no. 2.
11. Antuf'yev, V. M. Investigation of the effectiveness of differently shaped finned surfaces in crossflow. Teploenergetika, 1965, no. 1.
12. Timofeyev, V. M. and Karasina, E. S. Heat transfer in the tube bundles of a cast-iron finned economizer. Izvestiya VTI (All-Union Heat Engineering Institute), 1952, no. 5.
13. Karasina, E. S. Heat transfer in bundles of tubes with circumferential fins. Izvestiya VTI, 1952, no. 12.
14. Kuznetsov, N. V. and Shcherbakov, A. Z. Experimental determination of heat transfer and aerodynamic drag of a cast-iron finned air preheater. Izvestiya VTI, 1951, no. 2.
15. Teplovoy raschet kotel'nykh agregatov (normativnyy metod) (Thermal design of boiler units (Standard Method)). GEI Press, Moscow–Leningrad, 1957.

216 REFERENCES

16. Kuznetsov, N. V. Rabochiye protsessy i voprosy usovershensvovaniya konvektivnykh poverkhnostey kotel'nykh agregatov (Working processes and questions of refinement of convective surfaces of boiler units). GEI Press, Moscow–Leningrad, 1958.
17. Brauer, H. Wärme und Strömungstechnische Untersuchungen an quer angeströmten Rippenrohrbündeln. Chem.-Ing. Technik, 1961, 33, H.5, H.6.
18. Stasiulevičius, J. K. Samoška, P. S., Skrinska, A. J., and Survila, V. J. Thermophysical studies of a staggered bundle of smooth tubes in crossflow of compressed air. Trudy Akad. Nauk Lit. SSSR, 1963, series B, 4(63).
19. Stasiulevičius, J. K. and Samoška, P. S. Heat transfer from bundles of smooth tubes in crossflow of air at high Re. Ibid.
20. Stasiulevičius, J. K. and Samoška, P. S. Heat transfer and aerodynamics of staggered bundles of tubes in crossflow of air at Re > 10^5. Inzh.-fiz. zhurn., 1964, v. 7, no. 11.
21. VDI-Warmeatlas, VDI-Verlag, Düsseldorf, 1957.
22. Barbarich, M. V. and Kirpichnikov, F. P. Novyye metody poperechnoy i poperechnovintovoy prokatki metallov (New techniques of transverse and transverse-helical rolling of metals). VNITI AN SSSR (All-Union Research Institute of Pipes of the USSR Academy of Sciences Press), Moscow, 1957.
24. Fastovskiy, V. G. and Petrovskiy, Yu. V. Sovremennye effektivnyye teploobmenniki (Modern effective heat exchangers). GEI Press, Moscow-Leningrad, 1962.
25. Kuntysh, V. B. and Iokhvedov, F. M. Heat transfer and aerodynamic drag of bundles of tubes with slit transverse fins. Kholodil'naya tekhnika, 1968, no. 6.
26. Fortescue, R. W. and Hall, W. Investigation of heat transfer from fuel elements. In; Voprosy yadernokh energetiki (Problems of nuclear energetics). 1957 (Russian translation), no. 6.
27. Ueda, T. and Harade, I. Experiment of heat transfer on surfaces with transverse fins for flow direction. Bulletin of ISME, 1964, v. 7, no. 28.
28. Yudin, V. F. Effect of thermal conductivity of fin metal and coolant on heat transfer from finned tubes in crossflow. Teploenergetika (in print).
29. Yudin, V. F., Tokhtarova, L. S. and Andreyev, P. A. Heat transfer and drag of staggered bundles with different heights and pitches of fins. Trudy TsKTI (Polzunov Boiler and Turbine Institute), 1966, no. 73.
30. Berman, Ya. A. Investigation and comparison of finned tubular heat transfer surfaces over a wide range of Re. Khimicheskoye i neftyanoye mashinostroyeniye, 1965, nos. 10 and 11.
31. Brauer, H. Spiralrippenrohre für Querstrom-Wärmeaustauscher. Kaltetechnik, 1961, v. 13, H. 8.
32. Žukauskas, A., Makariavičius, V. and Slančiauskas, A. Teplootdacha puchkov trub v poperechnom potoke zhidkosti (Heat transfer from bundles of tubes in crossflow). Mintis Press, Vil'nyus, 1968.
33. Jameson, S. L. Tube spacing in finned tube banks. Trans. ASME, 1945, v. 67, no. 8.
34. Zozulya, N. V., Khavin, A. A. and Kalinin, B. A. Information letter No. 54. Thermal Energetics Institute of the Ukranian Academy of Sciences, 1962.
35. Yudin, V. F. and Tokhtarova, L. S. Investigation of heat transfer and drag of finned staggered bundles with different fin shapes. Energomashinostroyeniye. 1964, no. 12.
36. Mirkovič, Z. Heat transfer and flow resistance correlation of helically finned tubes in crossflow of staggered tube banks, 1972. Intern. Seminar, Trogir, Yugoslavia, 1972. In: Heat exchangers: design and theory source. Hemisphere Publishing, 1974.
37. Schmidt, Th. E. Der Wärmeübergang an Rippenrohre und die Berechnung von Rohrbündel-Wärmeaustauschern. Kältetechnik, 1963, v. 15, H. 4.
38. Yudin, V. F., Tokhtarova, L. S. Lokshin, V. A. and Tulin, S. N. Correlation of experimental data on convective heat transfer in crossflow over bundles with transverse spiral and circumferential fins. Trudy TsKTI, 1968, no. 82.
39. Krischer, O. and Kast, W. VDI-Forschungsheft. 1959, no. 474.

REFERENCES 217

40. Grass, G. and Coenen, F. P. Systematische Untersuchungen über den Wärmeübergang und Strömungswiderstand von Rippenrohren. Atomkernenergie, 1959, H. 2.
41. Lapin, A. and Schurig, W. F. Heat transfer coefficients for finned exchangers. Industr. and Engng. Chem., 1959, v. 51, no. 8.
42. Hirschberg, H. G. Wärmeübertragung und Druckverlust an quer angeströmten Rhorbuñdeln. Abh. des D. Kältetechn. Vereins, 1961, H. 16.
43. Hufschmidt, W. Die Eigenschaften von Rippenrohr-luftkühlern im Arbeits-bereich der Klimaanlagen. Forschungsber, Nordrhein-Westfalen, 1969, N 889.
44. Kast, W. Wärmeübergang an Rippenrohrbündeln. Chem.-Ing. Technik, 1962, v. 34, H. 8.
45. Migay, V. K. Effect of nonuniformity of heat transfer over the height of a fin on its effectiveness. Inzh-fiz. zhurn., 1963, no. 3.
46. Brauer, H. Wärmeübergang und Strömungswiderstand bei fluchtend und versetzt angeordneten Rippenrohren. Mannesmann, Forschungsber, 1962, H. 154.
47. Sasin, V. I. Effectiveness of finned surfaces of plate-type air cooler. Kholodil'naya tekhnika. 1965. no. 3.
48. Skrinska, A. J. and Stasiulevičius, J. K. Experimental study of the effect of nonuniformity of the heat transfer coefficient on the effectiveness of finned tubes. Trudy Akad, Nauk LitSSR, 1965, series B, 1(40).
49. Žukauskas, A., Stasiulevičius, I. and Skrinska, A. Experimental investigation of heat transfer of a tube with spiral fins in crossflow. In: Proc. 3rd Internat. Heat Transfer Conference, III, Chicago, 1966.
50. Reporter's Remarks, Author's Rebuttal. In: Proc. 3rd Internat. Heat Transf. Conference, VI, Chicago, 1966.
51. Lymer, A. and Ridal, B. F. Finned tubes in a crossflow of gas. The Journal of the Brit. Nucl. Energ. Conf, 1961, v. 6, no. 6.
52. Pshenisnov, I. F. and Lukhnov, M. I. Investigation of the effect of nonuniformity of heat transfer over the surface of a circumferential fin on its effectiveness. Teploenergetika, 1970, no. 9.
53. Lemon, A. W., Colburn, A. P. and Nottage, H. B. Heat transfer from a baffled-finned cylinder to air. Trans. ASME, 1945, v. 67, no. 11.
54. MacAdams, W. H., Drexel, R. E. and Golday, R. H. Local coefficients of heat transfer for air flowing around a finned cylinder. Trans. ASME, 1945, v. 67, no. 11.
55. Weiner, I. H., Gross, D., Paschkis, V. An experimental determination of local boundary conductances for an unbaffled circular finned cylinder. In: Gen. discussion heat transfer. inst. mech. eng. 1951.
56. Comossa H. Zur Lösung stationärer Wärmeleitprobleme mit Hilfe von Analogie-verfahren. Dissertation, Karlsruhe, 1956.
57. Neal, S. B. and Hitchcock, I. A. A study of the heat transfer processes in banks of finned tubes in crossflow. In: Proc. 3rd internat. heat transfer conference, III, Chicago, 1966.
58. Skrinska, A. J., Žukauskas, A. A. and Stasiulevičius, J. K. Experimental study of local coefficients of heat transfer with spirally finned tubes. Trudy Akad. Nauk LitSSR, 1964, Series B, 4(39).
59. Aerodinamicheskiy raschet kotel'nykh ustanovok (Aerodynamic design of boiler units). Energiya Press, Moscow, 1964.
60. Yudin, V. F. and Tokhtarova, L. S. Heat transfer and drag of staggered bundles of tubes with transverse fins in crossflow. Teploenergetika, 1973, no. 2.
61. Antuf'yev, V. M. Comparative investigation of convective surfaces on the basis of energy parameters. Energomashinostroyeniye, 1964, no. 5.
62. Antuf'yev, V. M. Replies to Polinovskiy's Remarks. Ibid., 1963, no. 7.
63. Mitskevich, A. I. A method for estimating the effectiveness of convective heat transfer. Trudy TsKTI, 1967, no. 78.

REFERENCES

64. Brauer, H. Untersuchungen an Querstrom- Wärmeaustauschern mit verschiedenen Rohrformen. Mitt. Verein Grosskesselbesitzer, 1961, nos. 4 and 6.
65. Kirpichev, M. V. On the most advantageous shape of heating surfaces. Izv. ENINa im. G. M. Krzhizhanovskogo, 1944, no. 12.
66. Yudin, V. F. A technique for comparative estimates of convective heating surfaces. Energomashinostroyeniye, 1969, no. 5.
67. Stasiulevičius, J. K., Skrinska, A. J., Survila, V. J. and Samoščka, P. S. heat transfer and drag of bundles of finned tubes in crossflow. Trudy Akad. Nauk LitSSR, 1971, Series B, 4(67). (Engl. transl. Heat Transfer-Soviet Research, 1973, v. 5, no. 6, pp 19–33.)
68. Kern, D. Q. Process heat transfer. McGraw-Hill, New York, 1950.
69. Isachenko, V. P. Osipova, V. A. and Sukomel, A. S. Teploperedacha (Heat transfer). Energiya Press, Moscow-Leningrad, 1965.
70. Kawashimo, K. and Katayama, K. Heat conduction in spiral fins. In: Proc. 9th Japan Nat. Congr. for Appl. Mech., 1959.
71. Schmidt. Th. E. Verbesserte Methoden zur bestimmung des Wärmeaustausches an berippte Flächen. Kältetechnik, 1966, v. 18, H. 4.
72. Schneider, P. I. Conduction heat transfer. Addison-Wesley Publ. Co., Mass., 1955.
73. Vanichev, A. P. An approximate method for solving problems of heat conduction at variable properties. Izv. Akad. Nauk SSSTR, OTN, 1946, no. 12.
74. Disinberre, G. M. Effective and local surface coefficients in fin systems. Trans. ASME, 1948, v. 80, no. 7.
75. Eck, B. Technische Strömungslehre, 5 Aufl. Berlin-Göttingen, 1957.
76. Wegener, G. Der Wärmeübergang an Kühlrippen Beih. Gesundh. Ing., 1929, v. 1, H. 24.
77. Yudin, V. F. and Tokhtarova, L. S. Investigation of the correction factor ψ of the analytic value of effectiveness of a circumferential fin. Teploenergetika, 1973, no. 3.
78. Koschinke, H. Das wirtschaflichste Rippenrohr. Arch. Wärmewirtsch. und Dampfkesselwes, 1933, 14.
79. Grimison, E. D. Correlation and utilization of new data on flow resistance and heat transfer for crossflow of gasses over tube banks. Trans. ASME, 1937, v. 59, no. 7.
80. Hausen, H. Neue Gleichung für die Wärmeübertragung bei freier und erzwungener Strömung. Allg. Wärmetechn., 1959, v. 9, 75/79.
81. Krischer, O. and Loos, G. Wärme und Stoffaustausch bei erzwungenger Strömung and Körper verschiedener Form. Chemi-Ing. Technik, 1958, 30.
82. Petukhov, B. S. Opytnoye izucheniye protsessov teploperedachi (Experimental study of heat transfer). Gosenergoizdat Press, Moscow, 1952.
83. Mikheyev, M. A. Osnovy teploperedachi (Fundamentals of heat transfer). Gosenergoizdat Press, Moscow, 1956.
84. Vargaftik, N. B. Spravochnik po teplofizicheskim svoystvam gazov i zhidkostey (Handbook on the thermophysical properties of gases and liquids). Fizmatgiz Press, Moscow, 1963.
85. Preobrazhenskiy, V. P. Teplotekhnicheskiye izmereniya i pribory (Heat engineering measurements and instruments). Gosenergoizdat Press, Moscow-Leningrad, 1953.
86. Yudin, V. F. and Tokhtarova, L. S. Heat transfer and drag of staggered and in-line finned bundles. Energomashinostroyeniye, 1964, no. 1.
87. Yudin, V. F. and tokhtarova, L. S. Comparison of methods of complete and local thermal simulation. Ibid., 1970, no. 12.
88. Gröber, H., Erk, S. and Grigull, U. Fundamentals of heat transfer. McGraw-Hill. (Russian translation, 1958).
89. Žukauskas, A. Advances in heat transfer. 8, Academic Press, New York, London, 1972.
90. Mayinger, F. and Schad, O. Örtliche Wärmeübergangszahlen in quer angeströmten Stabbündeln. Wärme und Staffübertragung, 1968, Bd. 1.
91. Žukauskas, A. A., Ulinskas, R. V. and Dauetas, P. M. Local heat transfer from a tube in a bundle in critical flow. trudy Akad. Nauk LitSSR, 1973, Series B, 2(75).

92. Eckert, E. VDI-Forschungsheft, 1942, no. 416.
93. Tamonis, M. M. Laminar boundary layer on isothermal surfaces with a power-law velocity distribution at the outer edge of the layer. Trudy Akad. Nauk LitSSR, 1965, Series B, 3(42).
94. Steimle, F. Zusammenhang zwischen Wärmeübergang and Druckabfall turbulenter Strömungen. Verland C. F. Müller, Karlsruhe, 1970.
95. Antuf'yev, R. M. Effektivnost' razlichnykh form konvektivnykh poverkhnostey nagreva (Effectiveness of different shapes of convective heating surfaces). Znaniye Press, Moscow-Leningrad, 1966.
96. Schmidt, Th. E. Vergleichszahlen zur Bewertung von Wämeaustauschern. Kältetechnik, 1949, v. 11. H. 1.

INDEX

Aerodynamic drag:
   of bundles, 17
Air velocity, 7, 85
Aluminum, 69, 111
   fins, 6, 21
   tubes, 9, 11
Argon, 70
Augmentation of heat transfer, 1
Averaging, 144
Axial flow, 51

Bare tubes, 11, 13, 63
Bare tube bundles, 81
Base tube area, 9
Bessel equation, 24, 26
Bessel functions, 25, 26
Bimetal tubes, 9
Biot number, 21
Blasius law, 67
Boundary layer, 8, 12, 63, 142, 149
   thickness, 51, 63, 64
Brass tubes, 9
Bundle:
   compactness, 162
   configuration, 16, 18
   drag, 14
   effectiveness, 153, 155

Calculating heat transfer, 22
Calculation of pressure drop, 180

Calorimeter, 82
   tubes, 114
Carbon, 69, 111
Carbon dioxide, 70
Cast iron fins, 7
Cast tubes, 7
Circular fins, 6–9, 12, 14, 15, 18, 20
   studies and development, 2
Circular surface, schematic of, 151, 152
Circular tube, 60
Circumferential distributing, 143
Circumferential fin, 23, 28, 33
   heat dissipation of, 27
   temperature gradient of, 22
Circumferential fins, 38, 60, 69
Conditional heat transfer coefficients, 5
Coefficient of heat transfer from a finned
   tube, 53
Constant fin geometry, 126
Constant thickness fin, schematic of, 23
Constants of integration, 25, 33
Continuous spiral fins, 9
   schematic of, 32
Convective heat transfer, 139
Convective heating surfaces, 154
Coolant flow, 4
Coolant thermal conductivities, 69
Cooling of tubes, 6
Copper, 69, 111
   fins, 21
   sleeve, 114
   tubes, 6, 9, 10

222 INDEX

Correlation of experimental data, 133
Crossflow, 2
Curvature of the spiral, 33
Cylinders, 71, 73
Cylindrical transverse fins, 108

Data on heat transfer, 171
Description of bundles, 79
Design recommendations, 176
Dimensions of bundles, 160
Distribution of heat 105, 106, 140
Drag, 2, 14
    coefficient values, 15

Euler number, 99–103, 174, 175
Experimental data analysis, 92
Experimental results, 115
Experimental techniques, 75
Experimental tube, schematic of, 114

Fin:
    base, 66
    —bearing tube, 16
    design, 19
    diameter, 7
    effectiveness, 34–36, 53, 58, 108, 112
    geometries, 6, 118
    geometry, 3, 14, 15
    height, 19, 50, 52, 55, 107, 147, 148, 165
    metal, 13
        thermal conductivities, 13
    parameters, 10
    pitch, 2, 5, 7, 16, 64, 146, 173
    profile, 22
    root diameter, 173
    shape, 50
    temperature, 19
    thermal conductivity, 3
    thickness, 5, 20, 31, 50, 61, 62, 91, 167
    tip, 3, 65, 142
    weight, 21
    with rectangular cross section, 32
Finned:
    bundles, 180
    surface, 35
    tube, 39, 53, 61, 30
        schematic of, 40, 58
        bundles, 50, 60, 90
        wall, 65

Finning:
    geometry, 12, 61, 78, 134, 163
    parameters, 66
Fins, 1, 2
    heat conduction of, 3
    rectangular, 3
    trapezoidally shaped, 3
Five-row bundle, 15
Flow pattern, 138
Flow:
    separation, 52
    temperature, 77, 91
    turbulence, 13, 142
    variables, 85
    velocity, 64
Fluid turbulence level, 9
Fourier series, 41
Free-stream velocity, 64
Friction drag, 14, 66

Gas flow in fins, 12
Gas heat exchanges, 1
Geometric parameters:
    of fins, 62
    of staggered bundles, 155

Heat conduction:
    characteristics, 38
    in a finned tube, 19
    of spiral fins, 31
Heat dissipation from a circumferential fin, 27
Heat distribution, 105, 106
Heat flux, 19, 21, 22, 30, 41, 45, 58
Heat fluxes, 1
Heating of tubes, 6
Heat transfer, 1, 2, 25, 38, 52
    augmentation, 1
    calculation, 2
    drag ratio, 52
    mechanisms, 12
    of finned tube bundles, 50
    rate, 8, 10, 17, 21
    schematics, 119–136
    surfaces, 1
Helium, 70
High thermal conductivity, 112
Hydraulic drag, 14, 18, 20, 87

INDEX 223

In-line bundles, 15, 62, 67, 69
In-line tubes, 13
Internal row, 67
Isotherms, 141

Laminar flow, 63, 73
    regions, 51
Laplace equation, 39
Linear reference division, 60, 61
Linear straight fins, 37
Local heat transfer, 82
Longitudinal fins, 38
Longitudinal bundle pitch, 127
Longitudinal rows, 15, 67–69, 74, 91
Low thermal conductivity, 112

Magnesium, 69
Mean heat transfer, 82
    coefficient, 118
Mean tube wall temperature, 45
Measuring calorimeter tube, schematic of, 83
Measuring local heat transfer, 84
Minimum flow cross sections, 7
Mixed flow zone, 177
Model heating methods, 149

Nomogram, 178
    for determining pressure drop, 181
    of kinematic viscosity, 89
Nomograms, 8
Nonuniformity of heat transfer, 108
Normal heat transfer, 34
Nusselt number, 21, 81

Optimal fin dimensions, 6, 10
Optimal fin shape, 21
Optimal finning parameters, 11

Planar flow in fins, 12
Plane rectangular fin, 20
Plates, 71
Pressure drop, 14, 153, 158
    in tube bundles, 92
    nomogram, 181
Principal parameters, selection of, 59
Prisms, 71, 73

Radial temperature gradient, 44
Radially averaged heat transfer, 144
Rate of heat transfer, 58
Ratio of flow velocities, 86
Rectangular fins, 3
Reduced heat transfer coefficient, 54
Reference velocity, 64
Resistance to heat transfer, 4
Reynold's equation, 153
Reynold's number, 8, 12, 17, 61, 75
Right circumferential fin, 140

Schematic of finned tube, 40, 58
Six-row bundle, 16
Smooth tube, 7
    bundles, 8, 91
Spheres, 71, 73
Spiral fins, 6, 8, 11, 15, 17, 18, 61, 112, 133
    heat conduction of, 31
Spirally finned system, 9
Spirally finned tubes, 11
Staggered bundles, 8, 10, 15, 62, 67
Staggered geometry, 73
Stainless steel, 69, 111
Steam-heated tubes, 9
Steel fins, 6
Stream velocity reduction, 62
Streamline length, 66
Steel tubes, 10
Surface friction, 91

Temperature:
    changes, 46
    distribution, 22, 141
    field in tube wall, 39
    field
        schematic of, 47
    fields, 50
    gradient, 23
        in a circumferential fin, 22
    isotherms, 46
Test data reduction, 81
Test loop, 76
Thermal conductivity, 38, 111
    over fin surface, 50
    schematic of, 70
Thermal effectiveness, 154
    of bundles, 157, 159
Thermal pattern, 142

Thermal performance, 63, 166
Thermal resistance, 23, 61
  of tube, 39
Thermal stimulation, 80
Thick tube wall, 45
Transverse bundle, 127
Transverse circumferential fins, 36
  graph of effectiveness, 37
Transverse rows, 91
Transversely finned tube, 4
Trapezoidal fin, 170
Trapezoidal fins, variation of thickness, 56
Trapezoidally shaped fins, 3, 64
Trapezoids, 28
Truncated triangles, 28
Tube:
  arrangements, 92
  bundle, 1
    arrangements, 5
    pressure drop in, 92
  configuration geometries, 65
  location, 14
  packing, 16
  pitch, 77
  rows, 67
  wall thickness, 59
Truncated triangles, 28
Turbulent flow, 63, 66
  regions, 51
Turbulent levels, 14

Variation of relative heat transfer, 145
VDI Warmeatlas handbook, 37
Velocities, 14
Velocity reduction, 12
Velocity variation, 51
von Karman equation, 63

Wall temperature, 9
Wedge-shaped vortex zone, 52
Weight of bundles, 160
Wind tunnel testing, 16